BIOTECHNOLOGY IN AGRICULTURE, INDUSTRY AND MEDICINE

CHITOSAN

MODIFICATIONS AND APPLICATIONS IN DOSAGE FORM DESIGN

BIOTECHNOLOGY IN AGRICULTURE, INDUSTRY AND MEDICINE

Additional books in this series can be found on Nova's website
under the Series tab.

Additional E-books in this series can be found on Nova's website
under the E-book tab.

BIOTECHNOLOGY IN AGRICULTURE, INDUSTRY AND MEDICINE

CHITOSAN

MODIFICATIONS AND APPLICATIONS IN DOSAGE FORM DESIGN

ASHOK KUMAR TIWARY
BHARTI SAPRA
GURPREET KAUR
AND
VIKAS RANA

Nova Science Publishers, Inc.
New York

For permission to use material from this book please contact us:
Telephone 631-231-7269; Fax 631-231-8175
Web Site: http://www.novapublishers.com

NOTICE TO THE READER

The Publisher has taken reasonable care in the preparation of this book, but makes no expressed or implied warranty of any kind and assumes no responsibility for any errors or omissions. No liability is assumed for incidental or consequential damages in connection with or arising out of information contained in this book. The Publisher shall not be liable for any special, consequential, or exemplary damages resulting, in whole or in part, from the readers' use of, or reliance upon, this material.

Independent verification should be sought for any data, advice or recommendations contained in this book. In addition, no responsibility is assumed by the publisher for any injury and/or damage to persons or property arising from any methods, products, instructions, ideas or otherwise contained in this publication.

This publication is designed to provide accurate and authoritative information with regard to the subject matter covered herein. It is sold with the clear understanding that the Publisher is not engaged in rendering legal or any other professional services. If legal or any other expert assistance is required, the services of a competent person should be sought. FROM A DECLARATION OF PARTICIPANTS JOINTLY ADOPTED BY A COMMITTEE OF THE AMERICAN BAR ASSOCIATION AND A COMMITTEE OF PUBLISHERS.

LIBRARY OF CONGRESS CATALOGING-IN-PUBLICATION DATA

Chitosan : modifications and applications in dosage form design / Ashok Kumar Tiwary ... [et al.].
   p. ; cm.
Includes bibliographical references and index.
ISBN 978-1-61761-174-2 (softcover)
1. Chitosan--Biotechnology. 2. Drugs--Controlled release. I. Tiwary, Ashok Kumar.
[DNLM: 1. Chitosan. 2. Dosage Forms. 3. Drug Design. QU 83]
TP248.65.C55C555 2010
615'.19--dc22
              2010029764

Published by Nova Science Publishers, Inc. † *New York*

CONTENTS

PREFACE

Chitosan is the deacetylated form of chitin. Generally, the substance becomes soluble in dilute acids when the degree of deacetylation is more than 50%. The solubility of chitosan in dilute acids is often needed to be modified when specific drug release properties have to be tailored into the dosage form. Chitosan carries free amine functionalities on the deacetylated units and hydroxyl groups on the acetylated as well as deacetylated units. Derivatization by introducing small functional groups such as, alkyl or carboxymethyl groups can increase the solubility of chitosan at neutral and alkaline pH without affecting its cationic character. In addition, chitosan can be grafted with other molecules through covalent binding. The amino groups can be used for acetylation, quaternization, reactions with aldehydes and ketones, chelation of metals etc. The hydroxyl groups can lend to o-acetylation, H-bonding with polar atoms etc. Primary derivatization followed by grafting improves the solubility, antibacterial, antioxidant, chelating, complexing, bacteriostatic and adsorbing properties while maintaining its mucoadhesivity, biodegradability and biocompatibility. These functionalities can also be used for interaction of chitosan with ions , polymers and drugs for obtaining materials useful for various applications. In its putative form, chitosan's positive charge allows it to interact with macromolecules like exogenous nucleic acids, negatively charged mucosal surfaces and plasma membranes. It is used as a support for gene delivery and cell culture. In addition, chitosan is reported to possess antibacterial, anti-fungal, anti-viral, anti-acid and anti-ulcer properties. The ability of chitosan to swirl across the membrane lipids is reported to be associated with its property to perturb the paracellular pathway for enhancing the permeation of hydrophilic drugs and to act as tight junction opener. This contributes to its role as a percutaneous and intestinal absorption enhancer. Chitosan derivative (*N,O*-carboxymethyl chitosan) and alginate blended with genipin was developed for site-specific protein drug delivery in the intestine that requires protection of drug release in the gastric pH. Tablets, microspheres, nanoparticles, intradermal vaccines etc. are being formulated using chitosan for modifying the drug release characteristics.

The available toxicological data on chitosan and its modified forms appears to indicate its safety for oral use because high doses have been found to be tolerated well in rodents and rabbits. However, its local action as a haemostatic together with its ability to activate macrophages and cause cytokine release may require a careful assessment of its safety for parenteral use.

Chapter 1

WHY CHITOSAN DERIVATIVES

The accumulated information about the physicochemical and biological properties of chitosan led to the recognition of this biopolymer as a promising biomaterial. Chitosan is a partially deacetylated polymer of N-acetyl glucosamine obtained after alkaline deacetylation of chitin, which is found in the shells of crustaceans, the exoskeletons of insects, and the cell walls of fungi. Chitosan, an unbranched cationic biopolymer, has three types of reactive functional groups that allow further chemical modification, i.e., amino or amido groups at C-2 positions as well as both primary and secondary hydroxyl groups at C-6 and C-3 positions, respectively (Fig. 1).

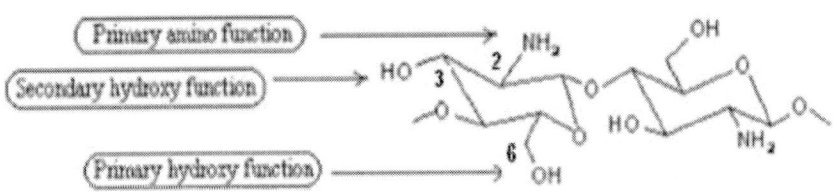

Figure 1. Structure of chitosan.

It displays interesting properties such as biocompatibility and biodegradability [Felt et al., 1998; Kumar et al., 2003], and its degradation products are nontoxic, nonimmunogenic, and noncarcinogenic [Bersch et al., 1995; Muzzarelli, 1997]. Chitosan is insoluble in water, aqueous alkaline solutions, and common organic solvents, but it readily dissolves in aqueous inorganic and organic acid media. Therefore, special attention has been paid to the chemical modification of chitosan. Cross-linking and graft copolymerization are well-known methods for the modification of chitosan and represent convenient and effective ways for improving the physical and mechanical properties for practical uses. Due to its polymeric character, chitosan has been used in the development of drug delivery systems [Prabaharan and Mano, 2005].

In all the studies enhanced absorption of chitosan has been found only in acidic environments in which the pH was less or of the order of the pKa value of chitosan (5.5–6.5). Chitosan, a weak base, requires a certain amount of acid to transform the glucosamine units into the positively charged water-soluble form. Due to the loss of charge in neutral and basic environments, chitosan precipitates from solution rendering it unsuitable as an absorption enhancer. At this pH, the molecule is most likely to exist in a coiled configuration [Artursson et al., 1997].

Further, the apical incubation of caco-2 cell monolayers with chitosan hydrochloride or chitosan glutamate at a pH of 6.2 resulted in reduction of Transepithelial Electrical resistance. At this pH, both chitosan salts (chitosan hydrochloride and chitosan glutamate) did not form clear solutions. In agreement with the results of the TEER experiments, no increase in the transport of the hydrophilic model compound [14C]-mannitol was observed at pH of 6.2 and 7.4 after incubation with these chitosan salts [Kotze et al., 1999]. The peroral route is considered to be the most convenient way of drug application for the patient. Most macromolecular pharmaceuticals such as peptide and protein drugs are indicated for chronic administration and therefore the peroral route will be the most suitable way of administering these drugs. The potential use of chitosans, as absorption enhancer in the more basic environments of the large intestine and colon, are limited.

In this regard it would be worthy to consider chitosan derivatives with different physicochemical properties, especially water solubility at neutral and basic pH values as they might prove to be useful as absorption enhancers in these environments. It was hypothesized by Kotze′ et al. (1999a) that polymers derivatives with different substituents, different basicities, or different charged densities could have the same or even increased efficacy in opening tight junctions than unmodified chitosan, which has a primary amino group. Chitosan is a versatile polymer with many functional groups available for chemical modification. Physicochemical and biological properties of chitosan are summarized in Table1.

Table 1. Physicochemical and biological proprieties of chitosan (Gupta and Ravi Kumar, 2000; Dutta et al., 2004)

Physicochemical properties of chitosan	Biological properties of chitosan
Molecular weight (MW) of 104 Da.	Biocompatible, Biodegradable
Possess three types of reactive functional groups; an amino group as well as both primary and secondary hydroxyl groups.	Haemostatic, bacteriostatic, and fungistatic
After heating decompose prior to melting, thus has no melt points.	Antitumor
Nearly all aqueous acids dissolve chitosan; most commonly used are formic acid and acetic acid.	Anticholesteremic
Multitude of cationic sites formed due to protonation of amino groups by acids along the chitosan chain increases its solubility by increasing both the polarity and the degree of electrostatic repulsion.	Safe and nontoxic
Forms salts with organic and inorganic acids.	Spermicidal
When protonated, adheres to negatively charged surfaces (muco-adhesive and forms gels with polyanions).	Mucoadhesive
High charge density at pH < 6.5.	Binds to mammalian cells aggressively
High molecular weight linear polyelectrolyte.	Regenerative effect on connective gum tissue
Aminable to chemical modification.	Accelerates bone formation
Viscosity, high to low.	Immunoadjuvant
Chelates certain transitional metals.	

Chapter 2

CHITOSAN FUNCTIONALIZATION

THIOLATED CHITOSAN

Thiol-containing chitosan, also called thiolated chitosan, is obtained through the reaction between chitosan and thiolactic acid. In this reaction, carbodiimide, N-(3-dimethylaminopropyl)-N'-ethylcarbodiimide hydrochloride (EDC) can be used to graft these two materials. EDC is a water-soluble carbodiimide that is typically employed in the 4.0–6.0 pH range. It is a zero-length crosslinking agent that has been widely used to couple carboxylic acid groups to primary amines. Thiolactic acid gets covalently attached to the primary amino group of chitosan. The carboxylic acid moieties of thiolactic acid are activated by EDC forming a O-acylurea derivative as an intermediate product that reacts with the primary amino groups of chitosan [Jayakumar et al., 2007].

The primary amino group at the 2-position of the glucosamine subunits of chitosan is the main target for the immobilization of thiol groups. Sulfhydryl bearing agents can be covalently attached to this primary amino group via the formation of amide or amidine bonds. In the case of the formation of amide bonds the carboxylic acid group of the ligands, cysteine and thioglycolic acid react with the primary amino group of chitosan mediated by a water soluble carbodiimide. The formation of disulfide bonds by air oxidation during the synthesis is avoided by performing the process at a pH below 5. At this pH-range the concentration of thiolate anions, representing the reactive form for oxidation of thiol groups, is low, and the formation of disulfide bonds can be almost excluded. Alternatively, the coupling reaction can be performed under inert conditions. For formation of amidine bonds, 2-iminothiolane is used as a coupling reagent. It offers the advantage of a simple one step coupling reaction. Investigations with all these thiolated chitosans showed that a degree of modification of 25–250 mmol thiol groups per gram of chitosan leads to the highest improvement in the mucoadhesive and permeation enhancing properties [Leitner et al., 2003]. Table 2 describes the preparation of thiolated chitosan derivatives along with their uses.

The improved mucoadhesive properties of thiolated chitosans are explained by the formation of covalent bonds between thiol groups of the polymer and cysteine – rich sub domains of glycoproteins in the mucus layer [Andreas et al., 2003; Leitner et al., 2003]. These covalent bonds are supposedly stronger than noncovalent bonds, such as ionic interactions of chitosan with anionic substructures of the mucus layer. This theory was supported by the results of tensile strength studies using tablets of thiolated chitosans, which demonstrated a positive correlation between the degree of modification with thiol bearing moieties and the adhesive properties of the polymer [Kast and Bernkop-Schnu¨rch, 2001]. These findings were confirmed by another *in vitro* mucoadhesion test system, where the time of adhesion of tablets on intestinal mucosa was determined. The contact time of the thiolated chitosan derivatives increased with increasing amount of immobilized thiol groups [Kast and Bernkop-Schnu¨rch, 2001; Kast and Andreas, 2001]. With chitosan-thioglycolic acid conjugates, a 5–10-fold increase in mucoadhesion in comparison to unmodified chitosan was achieved. The mucoadhesive properties of chitosan-4-thio-butyl-amidine (chitosan-TBA) conjugates were even better. This phenomenon could be attributed to additionally increased mucoadhesive properties due to improved ionic interactions between the additional cationic amidine substructure of the chitosan-TBA conjugate and anionic substructures of the mucus layer.

The permeation enhancing capabilities of chitosan for the first time were revealed by Illum et al., 1994. Chitosan was able to enhance the paracellular route of absorption, which is important for the transport of hydrophilic compounds such as therapeutic peptides and antisense oligonucleotides across the membrane. Various studies carried out on Caco-2 cell monolayers demonstrated a significant decrease in the TEER after the addition of chitosan [Artursson et al., 1997; Dodane et al., 1999; Martien et al., 2008]. The mechanism underlying this permeation enhancing effect seems to be due to the positive charges of the polymer, which interact with the cell membrane resulting in a structural reorganization of tight junction-associated proteins [Schipper et al., 1997]. In the presence of the mucus layer, however, this permeation enhancing effect is comparatively lower, as chitosan cannot reach the epithelium because of size limited diffusion and/or competitive charge interactions with mucins [Schipper et al., 1999]. Nevertheless, the results obtained on Caco-2 cell monolayers were confirmed by *in vivo* studies that revealed an enhanced intestinal absorption of the peptide drug buserelin in rats after co-administration with chitosan hydrochloride [Thanou et al., 2000]. The permeation enhancing effect of chitosan can be strongly improved by the immobilization of thiol groups. This effect of thiolated chitosans was shown in various permeation studies in Ussing type chambers using freshly excised intestinal mucosa. The uptake of fluorescence labeled bacitracin was improved 1.6-fold utilizing 0.5% of chitosan-cysteine conjugate instead of unmodified chitosan [Bernkop-Schnürch et al., 1999].

The uptake of the cationic marker compound rhodamine 123 was 3-fold higher in the presence of thiolated chitosan as compared to that in the presence of unmodified chitosans [Langoth et al., 2004]. This improved permeation enhancement has been ascribed to the inhibition of protein tyrosine phosphatase. This enzyme is believed to be involved in the opening and closing process of the tight junctions due to its involvement in the dephosphorylation of tyrosine subunits of occludin that represents an essential transmembrane protein of the tight junctions. When these tyrosine subunits of occludin are dephosphorylated, the tight junctions are closed. In contrast, when these tyrosine subunits are phosphorylated, the tight junctions are opened. The inhibition of protein tyrosine phosphatase by compounds such as phenylarsine oxide, pervanadate or reduced glutathione leads consequently to phosphorylation and opening of the tight junctions [Clausen et al., 2002; Barrett et al., 1999; Staddon et al., 1995]. In contrast to the stable but toxic protein tyrosine phosphatase inhibitors phenylarsine oxide and pervanadate, the inhibitory effect of glutathione is lower as it is rapidly oxidized on the cell surface loosing its inhibitory activity [Grafstrom et al., 1980]. However, in combination with thiolated chitosans, the oxidation of glutathione on the membrane is restricted because thiomers are capable of reducing oxidized glutathione [Clausen et al., 2002].

Table 2. Preparation of thiolated chitosan derivatives and their uses

Derivative/Modification	Reaction with	Example	Property	Application	References
Thiolated chitosan (chitosan–thioethylamidine)	Derivatization of primary amino groups of chitosan with coupling agents bearing thiol functions	Polycarbophil or thiomers e.g. chitosan-cysteine, chitosan-thioglycolic acid, chitosan-4-thio ethylamidine conjugate	Insufficient stability of thiomer	Mucoadhesive properties can further be improved by immobilization of thiol groups on the polymer. Improved permeation enhancing properties. Enhanced and sustained gene delivery. Due to excellent cohesive property, prolonged controlled release of embedded ingredients. In situ gelling or cross linking can be observed within a range of 5-6 pH therefore absorption of thiolated chitosan on vaginal, nasal, buccal and ocular mucus becomes possible. Enzyme inhibitory properties can be beneficial for oral administration of peptide and protein drugs.	Hassan and Gallo, 1990; Bernkop-Schnurch et al., 1999; Schipper et al., 1999; Bernkop-Schnurch and Hopf, 2001; Kast and Bernkop-Schnurch, 2001; Leitner et al., 2003; Kast et al., 2003; Guggi et al., 2002; Guggi et al., 2003; Krauland et al., 2004; Maculotti et al., 2005; Lee et al., 2006; 2007

Derivative/Modification	Reaction with	Example	Property	Application	References
	Imidoester reaction with isopropyl-S-acetylthioacetimidate	Chitosan-thioethylamidine derivative; chitosan–4-thio-butyl-amidine	By the immobilization of thiol groups mucoadhesion was strongly improved due to the formation of disulfide bonds with mucus glycoproteins.	New polymeric excipient for various drug delivery systems; improvement of both the zinc-binding capacity and the antiprotease effect of chitosan was observed when the biopolymers (chitosan and thiolated chitosan) were used as coating component of the core–shell nanoparticles, in comparison with their behaviour in solution.	Kafedjiiski et al., 2005; Bravo-Osuna et al., 2008
		Chitosan 2-iminothiolane conjugate	Improved mucoadhesive and permeation enhancing properties, more stable in aqueous media with respect to unmodified chitosan	Promising tool for the non-invasive administration of hydrophilic macromolecular drugs.	Maculotti

The chemical modification of chitosan via derivatization with various reagents bearing sulfhydryl groups causes a dramatic change in the polymer's properties. Mucoadhesiveness and cohesiveness are strongly improved. A comparatively stronger permeation enhancing effect can be achieved, which can be further improved by combining thiolated chitosans with the permeation mediator glutathione. Furthermore, thiolated chitosans display in situ-gelling features and facilitate controlled drug release. Due to these advantageous features thiolated chitosans have been successfully used for peroral administration of peptide drugs. They seem to represent a promising new generation of polymeric excipients in particular for the delivery of hydrophilic macromolecular drugs. Thiolated chitosans might also prove successful as scaffold material in tissue engineering and as coating material for stents.

Chapter 3

TRI METHYLATED CHITOSAN

Tri methylated chitosan (TMC), a partially quaternised derivative of chitosan, has intensely been studied and described by Kotze´ et al. [1999a] for its absorption enhancing effects. It was concluded that the potential use of TMC, in neutral and basic environments where normal chitosan salts are ineffective as absorption enhancers, could contribute significantly to the effective delivery of hydrophilic compounds such as protein and peptide drugs. Recently, another derivative of chitosan, mono-N-carboxymethyl chitosan (MCC) was also evaluated for its absorption enhancing ability [Thanou et al., 2001b].

TMC is a partially quaternised derivative of chitosan which is prepared by reductive methylation of chitosan with methyl iodide in a strong basic environment at an elevated temperature (Table 3). The degree of quaternisation can be altered by increasing the number of reaction steps or by increasing the reaction time [Sieval et al., 1998; Snyman et al., 2002]. The chitosan used for synthesizing TMC was only soluble in acidic solutions, but after quaternisation it became perfectly soluble in water (Figure 1). This enhanced solubility, either in basic or acidic medium, was observed for low degree of quaternisation of 10% [Kotze´ et al., 1998]. TMC exhibited superior solubility and basicity, even at low degrees of quaternisation as compared to chitosan salts. The increase in solubility was attributed to the replacement of the primary amino group on the C-2 position of chitosan with quaternary amino groups. It should be noted that the molecular weight of the polymer chain increases during the reductive methylation process due to the addition of methyl groups to the amino group of the repeating monomers. However, a net decrease in the absolute molecular weight was observed due to degradation of the polymer chain caused by exposure to the specific reaction conditions during the synthesis [Snyman et al., 2002]. The mucoadhesive properties of TMC with different degrees of quaternisation, ranging between 22 to 49%, were investigated by Snyman et al [2003]. TMC was found to have a lower intrinsic mucoadhesivity as compared to the chitosan salts, chitosan hydrochloride and chitosan glutamate. However, as compared to the reference polymer pectin, TMC possessed superior mucoadhesive properties. The decrease in the mucoadhesivity of TMC could be explained by a change in the conformation of the TMC polymers due to interactions between the fixed positive charges on the quaternary amino groups, which possibly also decreased the flexibility of the polymer molecules. The interpenetration into the mucus layer by the polymer was influenced by a decrease in flexibility consequently resulting in decreased mucoadhesivity [Snyman et al., 2003].

Incubation of intestinal epithelial cells (Caco-2) with TMC (1.5, 2.0 or 2.5% w/v with a degree of quaternisation of 12% resulted in an immediate reduction in TEER values. The reduction in TEER was 9 ± 4, 52 ± 3 and $79 \pm 0.3\%$, respectively, after 20 min. Prolonged incubation only resulted in a gradual decrease in resistance compared with the initial reduction in TEER after 20 min. The highest reduction in TEER was measured at a concentration of 2.5% w/v thus indicating that the reduction in TEER was concentration dependent [Schipper et al., 1997]. With removal of the polymer solutions, repeated washing and substitution of the apical medium with fresh Dulbecco's Modified Eagles Medium, reversibility of the effects was noticed, especially at 1.5 and 2.0% w/v concentrations of TMC. The decrease in TEER 20 min after incubation, at 0.25% w/v concentration, followed the order chitosan hydrochloride ($71 \pm 4\%$ reduction), chitosan glutamate ($56 \pm 1\%$ reduction), TMC ($28 \pm 1\%$ reduction), suggesting that the chitosan salts were more effective than TMC at similar weight concentrations [Kotze' et al., 1998].

Table 3. Quaternized chitosan derivatives and their applications

Derivative/Modification	Reaction with	Example	Property	Application	References
Quaternized chitosan Quaternized chitosan	Methyl iodide in alkaline solution of N-methyl pyrrolidone	N,N,N-trimethyl chiosan chloride	Much higher aqueous solubility in much broader pH and concentration range	Absorption enhancer Improvement of mucoadhesive property, antibacterial activity	Kim et al., 2002; Merwe et al., 2004; Atyabi et al., 2007; Sajomsang et al., 2009
	Cross linked with glutaradehyde	Cross linked N-Trimethylated Chitosan		Gene carrier in epithelial call line Microsphere fabrication	Sieval et al., 1998; Thanou et al, 1999; 2000a; 2002; Kean et al., 2005
		N-octyl- N-trimethyl choitosan derivatives	Increased solubility	Polymeric micelle formation, solubilization and controlled release of 10-hydroxycamptothecin	Peng and Zhang, 2005
N,N,N-Trimethylated chitosan	Schiff's reaction with aldehydes/ketones	N-alkyl derivative (quaternisation)		Antibacterial activity	Jia et al., 2001; Avadi et al., 2004
		N,N,N-Trimethyl Chitosan			
		N-propyl- N,N-dimethyl chitosan			

Table 3. Continued

Derivative/Modification	Reaction with	Example	Property	Application	References
N-Aryl chitosans		N-furfuryl- N,N-dimethyl chitosan			
		N- diethyl-methyl amino chitosan			
N,N,N-trimethylammonium chitosan chloride (TMChC)	Methylation	Quaternised alkyl chitosan	Improved polycationic properties	Antifungal	
	Salicylaldehyde derivatization	N-arylidene chitosan		Antioxidant similar to chitosan	
		Chitosan-schiff's base reaction with methoxy phenyl aldehyde e.g. vanillin, o– vanillin, syringaldehyde, veratraldehyde	Impart insolubility to chitosan, biodegradable and mechanically resistant		
The methylated N-aryl chitosan derivative ;methylated N-(4-N,N-dimethylaminocinnamyl) chitosan chloride)	Reductive alkylation with octyl, decyl and dodecyl aldehydes	N,N dialkyl chitosan		Vesicles (for drug release experiment)	Li and Xin, 2006; Li et al., 2007
	With alkyl halides under basic conditions	Alkyl Chitosan e.g. isobutyl chitosan	Improved solubility in neutral aqueous solution due to reduction in	DNA delivery (dodecyl chitosan)	Liu et al., 2003

Method	Derivative	Property	Application	Reference
	N-Carboxybenzyl, glycine-glucan, *N*-carboxy-(aryl) chitosans methyl chitosan,	crystallinity of chitosan	Chromatographic media and metal ion collection	Gupta and Kumar, 2000
Folate conjugated	*N*-trimethyl chitosan (folate-TMC)		The enhancement of cellular uptake. Nanoparticles were promising carrier for protein	Zheng et al., 2009
By reacting *N*-chloroacyl-6-*O*-triphenylmethylchitosans with tertiary amines.	N-(1-Carboxymethyl-2-pyridinium)chitosan chloride, N-[1-Carboxymethyl-2-(1-methylimidazolium)]chitosan chloride, N-(1-Carboxymethyl-2-tributylammonium)chitosan chloride	Water-soluble quaternary chitosan derivative, Full degree of substitution of the quaternization step was obtained		Holappa, et al., 2006

Table 4. In vitro studies of TMC derivatives as absorption enhancers

TMC Derivative	Degree of Quaternization (%)	Drug	Enhancement	References
TMC	12.28	Buserelin	28-73 fold	Kotzé et al., 1997a
		FITC-Dextran	167-373 fold	Kotzé et al., 1997a
		Insulin	Increased transport	Kotzé et al., 1997b
		Mannitol	Increased transport	Kotzé et al., 1998
		PEG 4000	Increased transport	Kotzé et al., 1998
	60	Octreotide acetate	Increased PC (34-121 fold)	Thanou et al., 2000a
	12-59	Mannitol	PC Increased with increase in degree of quaternization (max. with TMC 49)	Hamman et al., 2003
TMC-H	61.2	Mannitol	Increased PC (31-48 fold)	Kotzé et al., 1999c
	19.9	Mannitol	Increased PC (97 fold)	Kotzé et al., 1999c
TMC-L	12.3	Mannitol	Increased PC	Kotzé et al., 1999c
	12.6	Mannitol	Increased PC (51 fold)	Kotzé et al., 1999c
TMC 40	40	Mannitol	Increased PC	Thanou et al., 2000b
		Buserelin	Increased PC (21 fold)	Thanou et al., 2000c
TMC 60	60	Mannitol	Increased PC	Thanou et al., 2000b
		Buserelin	Increased PC (60 fold)	Thanou et al., 2000c
TMC 22	22	Mannitol	PC Increased with increase in degree of quaternization (max. with TMC 49)	Hamman et al., 2003
TMC 38	38			
TMC 43	43			
TMC 49	49			

*PC: Partition coefficient.

Exposure of the apical side of the caco-2 cells monolayers to 0.25% w/v of the polymers resulted in a 34-fold (chitosan hydrochloride), 25-fold (chitosan glutamate) and 11-fold (TMC) increase in the absorption rate of

[^{14}C]-mannitol, compared to the control group. Similar results were obtained for [^{14}C] PEG 4000. At higher concentrations, TMC was able to increase the partition coefficient and transport values further for both [^{14}C]-mannitol and [^{14}C] PEG 4000. An increase of 17- and 21-fold in transport values were observed for [^{14}C]- mannitol at 2.0 and 2.5% w/v concentrations of TMC, respectively. The same tendency was observed with [^{14}C] PEG 4000 [Kotze' et al., 1998]. From these results, it was evident that TMC was not as effective at similar weight concentrations as chitosan hydrochloride and chitosan glutamate. Further, TMC with a degree of quaternisation of 12% was found to increase the transport of fluorescein isothiocyanate-labelled dextran (FD-4) across Caco-2 cell monolayers. The transport of this large hydrophilic model compound (4400 Da) was increased ~ 167, ~ 274- and ~ 373-fold with 1.5, 2.0 and 2.5% w/v concentrations of TMC, respectively [Kotze' et al., 1997a].

TMC was also able to increase the transport of insulin at a pH of 4.4 compared to the control where no transport was observed. The transport of insulin was increased to 0.3% and 0.8% of the total dose applied at 1.5% and 2.5% w/v concentrations of TMC, respectively. An increase in the transport of buserelin (1300 Da) was also observed at a pH of 6.2. The transport was increased to 1.4% and 2.7% of the total dose applied with 1.5 and 2.5% w/v solutions of TMC, respectively, as compared to the control group where only 0.04% of total dose applied was transported [Kotze' et al., 1997a]. The different *in vitro* studies performed with TMC are summarized in (Table 4).

The different variables that can influence the enhancing properties of these polymers have been included. TMC derivatives are especially effective in enhancing the transport of small hydrophilic compounds (e.g., mannitol), though they also improve the transport of large molecules such as buserelin, insulin, and octreotide acetate.

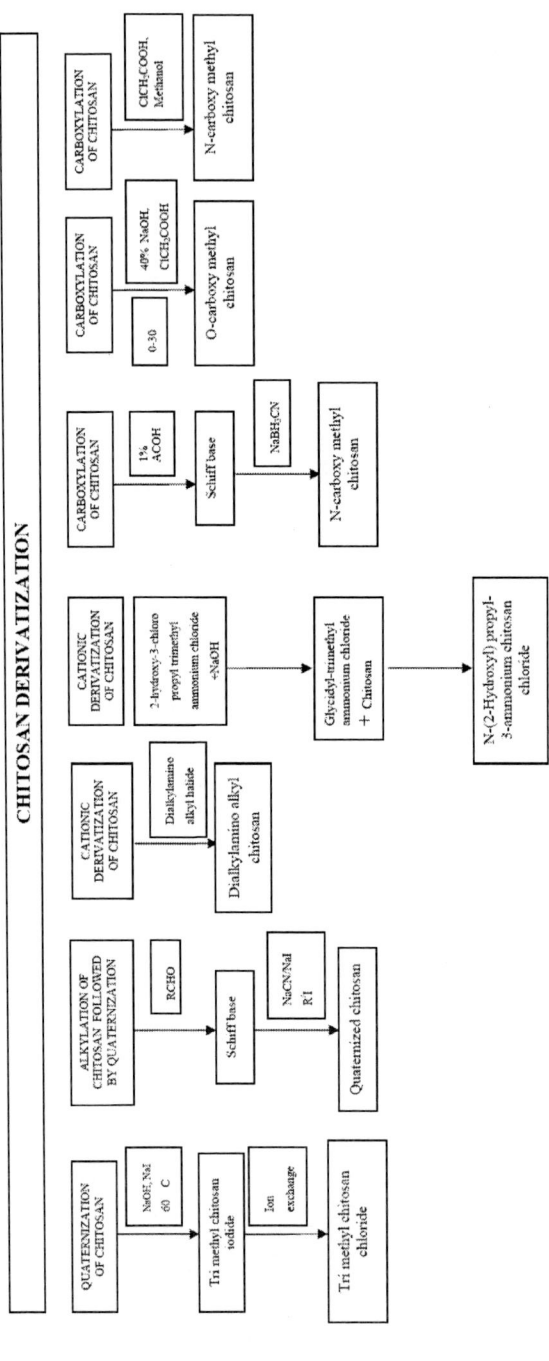

Figure 1. Approaches for preparing chitosan derivatives.

Chapter 4

HIGHLY CATIONIC DERIVATIVES

Chitosan can be modified in different ways to introduce a cationic charge. Modifications, such as alkylation or acylation, can be executed at the amino function of chitosan to obtain a quaternized amino group carrying the cationic charge (Table 5). Alternatively, a moiety itself carrying cationic charge can be introduced. This moiety can be covalently linked to one of the chitosan functionalities such as the hydroxyl or the amino group. Modifications which result in enhanced permanent cationic charge on the chitosan molecule are highly preferred, as they exhibit a pH-independent positive charge. Depending on the magnitude of cationic charge introduced on the chitosan molecule, water solubility at neutral to basic pH values can be achieved, in contrast to unmodified chitosan which is soluble only under acidic conditions. A simple way to introduce a permanent cationic charge to the chitosan molecule is the methylation of chitosan's amino groups, resulting in N-trimethyl chitosan salts, notably the chloride form [Figure 1] [Sieval et al., 1998]. The cationic polymers are prepared by reaction of chitosan and dialkylaminoalkyl chloride in alkaline condition [Je et al., 2006]. Chitosan derivatives of dialkylaminoalkyl type with N-aminoethyl, N-diethylaminoethyl, N-dimethylaminoethyl, N-dimethylaminoisopropyl etc. have exhibited the ability to enhance the penetration of both hydrophobic and hydrophilic molecules across the excised porcine cheek epithelium (Table 5). This effect was observed to be stronger than that of the known absorption enhancer TMC [Zambito et al., 2006].

A water-soluble derivative of chitosan, N-(2-hydroxyl) propyl-3-trimethylammonium chitosan chloride (HPTCC) was synthesized by its reaction between glycidyl-trimethyl- ammonium chloride [Xu et al., 2003].

HPTCC was employed for making nanoparticles for protein delivery utilizing its ionic gelation with sodium tripolyphosphate [Xu et al., 2003]. Highly cationic chitosans find applications in cosmetics for hair and skin care too. Chitopearl products (Fuji Spinning Co., Japan) belong to the class of highly cationic derivatives of chitosan. They comprise of chitosan porous beads cross-linked by bifunctional reagents such a diisocyanate or diepoxy derivatives [Seo et al., 1989a]. Chitopearl spherical chitosan particles produced from diisocyanates are suitable for chromatographic purposes and as enzyme supports [Yoshida et al., 1994]. Treatment of chitin solution in solvents such as DMA–LiCl or anhydrous pyridine with excess 1,6-diisocyanatohexane and exposure to water vapor for two days produced flexible, opaque materials whose main characteristics included insolubility in aqueous and organic solvents, remarkable crystallinity, typical infrared spectrum and degree of substitution but no thermoplasticity. Microencapsulation of lactic acid bacteria based on the cross-linking of chitosan by 1,6-diisocyanatohexane has been performed [Grobouillot et al., 1993]. The cationic character of the chitosan is pivotal to many of its applications such as bioadhesion, absorption enhancement, transfection efficiency as well as biological activities such as antitumor, antimicrobial, antiinflammatory, and antihypercholesterolemic effect.

Table 5. Derivatization of chitosan for enhancing its cationic character

Derivative/Modification	Reaction with	Example	Property	Application	References
Highly cationic derivatives Dialkylaminoalkylchitosan	Amino alkylation (i.e. with dialkyl amino alkyl chloride in alkaline condition)	Partially substituted N,O-[(N,N- diethyl amino methyl diethyl di-methylene ammonium)] methyl chitosan		Enhanced penetration of hydrophobic/hydrophilic materials across porcine epithelium	Lee et al., 2002; Je and Kim, 2005; 2006;
	Glcydyl-tri-methyl ammonium chloride	N-(2-hydroxyl)propyl-3-trimethyl ammonium chitosan chloride		Nanoparticles for protein delivery (ionic gelation with sodium tripolyphosphate)	Xu et al., 2003
	Cross-linking with bifunctional reagents e.g. diisocyanate or diepoxy derivatives			Chitopearls for chromatographic purposes and as enzyme supports	
	DMA–LiCl or anhydrous pyridine with excess 1,6-diisocyanotohexane	Chitin	Flexible, opaque material insoluble in aqueous and organic solvents, remarkable crystallinity, typical IR but no thermoplasticity	Microencapsulation of lactic acid bacteria	Grobouillot et al., 1993
	Chitosan conjugated with polyethylenimine and methoxy poly(ethylene glycol).	Methoxy poly(ethylene glycol)–polyethylenimine–chitosan	Polycationic polyethylenimine was introduced onto chitosan for the purpose of enhancing the positive charge of the copolymer	Effective gene delivery and package molecule.	Xu et al., 2009

Chapter 5

HYDROXYALKYL CHITOSAN

Hydroxyalkyl chitosans are obtained on reacting chitosan with epoxides (ethylene oxide, propylene oxide, butylenes oxide) and glycidol (Table 6). Depending on the epoxide and conditions (e.g. solvent and temperature), the reaction may take place predominantly at the amino or alcohol group, yielding N-hydroxyalkyl chitosan or O-hydroxyalkyl chitosan or a mixture of both [Lang et al., 1988; 1989; 1990; Donges et al., 2000]. The choice of catalyst (NaOH or HCl) and reaction temperature determines the ratio of O/N-substitution (hydroxypropylation of chitosan by propylene oxide) [Maresh et al., 1989]. Hydroxypropylation of chitosan with propylene oxide is controlled by the pH of the solution. The substitution reaction occurred preferentially at the amino groups or hydroxyl groups under neutral or alkaline conditions, respectively [Maresch et al., 1989; Lang et al., 1997]. Hydroxyalkylations with 2-chloroethanol and 3-chloropropane-1,2-diol were also reported [Tokura et al., 1983]. Synthesis of hydroxypropylchitosan and its evaluation for use as an antimicrobial agent [Peng et al., 2005] or temperature sensitive injectable carrier for cells [Dang et al., 2006] has been carried out by different researchers. O-hydroxyethylchitosan (glycol chitosan) was synthesized by reaction with 2-chloroethanol in alkaline medium [Ronghua et al., 2003]. The epoxides used for hydroxyalkylation of chitosan can be substituted for example with carboxylic groups [Gruber et al., 1995; 1997].

Self-assembled nanoparticles based on glycol chitosan were prepared as a carrier for paclitaxel and doxorubicin [Kim et al., 2006; Park et al., 2006]. Glycol chitosan was also used as stabilizer for protein encapsulated into poly(lactide-co-glycolide) microparticle [Lee et al., 2007].

Table 6. Derivatization of hydroxyl alkyl chitosans

Derivative/ Modification	Reaction with	Example	Property	Application	References
Hydroxy alkyl chitosans	Epoxides (ethylene oxide, propylene oxide, butylene oxide) and glycidol	N-hydoxy alkyl chitosan	Products having marked surface activity and foam enhancing properties		Roberts et al., 1992
	2-chloroethanol in alkaline medium	O-hydoxy alkyl chitosan	Oligomers at a temperature higher than 40°C		Maresh et al., 1989
		Hydroxy propyl chitosan	Antimicrobial temperature sensitive injectable carrier for cells		Peng et al., 2005; Dang et al., 2006
		Glycol chitosan	Nanoparticles as carrier for paclitaxel, doxorubicin	As stabilizer for protein encapsulated into poly (lactide-co-glycolide) microparticles	Kim et al., 2006; Park et al., 2006; Lee et al., 2007

Chapter 6

CARBOXYALKYL CHITOSAN

The process of carboxyalkylation introduces acidic groups on the polymer backbone. On introduction of carboxyl groups on the amino groups of chitosan, amphoteric polyelectrolytes containing both cationic and anionic fixed charges are obtained (Table 7). By varying the degree of substitution of the carboxyl bearing group, various charge densities on the molecular chain can be obtained, which provide a convenient way to control pH-dependent behaviour. The other synthetic route that is selective in the formation of N-carboxyalkylation uses carboxyaldehydes in a reductive amination sequence [Muzzarelli et al., 1982]. By using glyoxylic acid, water-soluble N-carboxymethyl chitosan is obtained, which is soluble in water, has high viscosity, large hydrodynamic volume, film and gel-forming capabilities [Muzzarelli et al., 1994] (Table 7). Due to these properties it has emerged as an attractive option for its use in food products and cosmetics [Muzzarelli et al., 1988; Pavlov et al., 1998]. Carboxymethyl chitosan is used for developing different protein drug delivery systems as super porous hydrogels, pH-sensitive hydrogels and cross-linked hydrogels [Chen et al., 2004a; 2004b; Lin et al., 2005; Yin et al., 2007]. N,N-dicarboxymethyl chitosan possesses good chelating abilities and its chelate with calcium phosphate favour osteogenesis while promoting bone mineralization [Muzzarelli et al., 1998]. O-carboxymethyl chitosan exhibits antibacterial activity and modified adhesive properties. In addition, it exhibits enhanced blood compatibility [Zhu and Fan, 2005]. Carboxymethyl chitosan and modified carboxymethyl chitosan have been employed as a carrier for delivering drugs like gatifloxacin, camptothecin, ibuprofen, and adriamycin [Zhu et al., 2006; Liu et al., 2006; Liu et al., 2007a; Zhu et al., 2007]. N-carboxyalkyl derivative of chitosan was synthesized and tested for antioxidant and antimutagenic activity [Skorik et al., 2003; Kogan et al., 2004].

Quaternized carboxymethyl chitosan (QCMC) was used for obtaining carboxymethyl chitosan (CMC). N-quaternary ammonium group was introduced by the reaction of CMC with 2, 3-epoxypropyl trimethylammonium. Degree of substitution (DS) of CMC does not significantly affect the antimicrobial activity. However, the antimicrobial activity of QCMC was found to be enhanced with increase of their DS of quaternization or the decrease of their molecular weight. QCMC was complexed with calcium hydroxide. Animal experiment results indicated that QCMC could induce reparative dentine formation and showed better ability to induce dentin as compared to calcium hydroxide [Sun et al., 2006].

Microwave technique was successfully used for crosslinking CMC films for possible use in wound care applications [Wongpanit et al., 2005]. FT-IR spectroscopy studies indicated that crosslinking of microwave-treated CM-chitosan films involved the carboxylate and the amino groups. Pure CM-chitosan films appeared to be amorphous. Based on both direct and indirect cytotoxicity assays, microwave treated CM-chitosan films were not found to be cytocompatible. Despite their observed cytotoxicity, biological response of human fibroblasts for microwave treated CM-chitosan films appeared to be normal, as evidenced from the amount of protein synthesized. Lastly, it was observed that human fibroblasts adhered on the surface of microwave-treated CM-chitosan films, which indicated the potential of these films in wound healing.

CMCS can be obtained from the reaction of chloroacetic acid and chitosan in alkaline condition with easy building blocks for application in drug delivery, tissue engineering, and viscosupplementation [Chen et al., 2004a; Liang et al., 2004]. Because of cationic amine groups and anionic carboxyl groups, the swelling, drug permeation, and release properties of CMC could be controlled by pH change [Chen et al., 2004b]. The major problem encountered with this degradable polysaccharide was its high solubility in aqueous media [Chen and Park, 2003]. As a consequence, film coatings consisting of this polymer alone would be unable to effectively control the drug release. However, the incorporation of CMC in water-insoluble film-forming polymers, such as poly(estersulfone) (PES) or poly(vinyl alcohol) (PVA), could provide a promising alternative [Zhao et al., 2003; Wang et al., 2007].

Table7. Derivatization of chitosans into carboxyalkyl chitosans along with the properties

Derivative/Modification	Reaction with	Example	Property	Application	References
Carboxyalkyl chitosans	Monohalocarboxylic acids	N-carboxy alkylated chitosan	Introduction of carbonyl groups onto amino groups of chitosan		Kim et al., 1997; Liu et al., 2001
	N-carboxyalkylation with carboxyaldehydes in a reductive amination sequence	O-carboxyalkyl chitosan	Control pH dependent behavior		Muzzarelli et al., 1982
	Glycoxylic acid	N-carboxy methyl chitosan	Water soluble, high viscosity, large hydrodynamic volume, film and gel forming capabilities	Used in food products and cosmetics. Development of different protein drug delivery systems e.g. superporous hydrogels, pH sensitive hydrogels, cross-linked hydrogels	Muzzarelli et al., 1994; Pavlov et al., 1998; Chen et al., 2004a; Chen et al., 2004b; Lin et al., 2005; Yin et al., 2007
		N,N –dicarboxy methyl chitosan	Good chelating properties	Favored osteogenesis while promoting bone mineralization	Muzzarelli et al., 1998; Zhu and Fan, 2005

O-Carboxymethyl chitosan

Table7. (Continued).

Derivative/Modification	Reaction with	Example	Property	Application	References
		O-carboxy methyl chitosan		Antibacterial activity, modifies adhesive properties, enhanced chondrocyte adhesion; surface modification of vascular grafts, enhance blood compatibility; as carrier for delivering drugs such as gatifloxacin, campoththecin, ibuprofen, adriamycin	Zhu et al., 2006; 2007; Liu et al., 2006; 2007a
		o-Carboxymethyl, cross-linked o-carboxymethyl		Molecular sieves, viscosity builders, and metal ion collection	Gupta and Ravi Kumar, 2000
	3-halopropionic acids under mild alkaline conditions and ambient temperature	Higher homolog of carboxymethyl chitosan i.e. N-(2-carboxyethyl) chitosan		Antioxidant and antimutagenic activity	Skorik et al., 2003; Kogan et al., 2004
	Arylonitrile	Cyanoethyl chitosan			

Derivative/Modification	Reaction with	Example	Property	Application	References
	Ethyl acrylate in aqueous acidic medium	N-carboxyethyl ester (intermediate)	Can be easily hydrolysed to free acid or used as intermediate to substitute with various hydrophilic amines, without requiring protecting groups		
	Reductive amination with 2-carboxy benzaldehyde and cross linked with glutaraldehyde	N-carboxy-benzyl chitosan	pH-sensitive hydrogel	Colon specific drug delivery of 5-FU	Lin et al., 2007
	Carboxyalkylation (with carboxyaldehydes)	4-hydroxyphenyl pyruvic acid	Stable self sustaining gels		
		Modified chitosan	Hydrolysed by lysozyme, lipase, papain		Kumar et al., 2004
		Chitosan-α-ketoglutaric acid and hydroxamated chitosan-α-ketoglutaric acid		Theophylline loaded iron (III)-cross-linked polymeric beads for prolonged drug release, as well as in augmenting adsorption properties	Ding et al., 2007a; 2007b

The CMC-coated capillary has been successfully applied to separate basic proteins and recombinant human erythropoietin (rhEPO). Furthermore, several experimental parameters, such as the concentration and pH of the running buffer, temperature, and applied voltage, were optimized for the separation of rhEPO glycoforms. Comparison of an uncoated capillary with chitosan- or CMC - coated capillaries for the separation of rhEPO glycoforms demonstrated that rhEPO glycoforms could be well separated by a CMC-coated capillary within 8 min with good reproducibility and resolution [Fu et al., 2007].

CMC injection spots did not show erythema or oedema after 24, 48 or 72 h suggesting complete hydrolysis within 3 days of implantation possibly due to the carboxymethyl groups in their structure. When carboxymethyl groups were synthesized with chitosan, the polysaccharide became an ampholyte, which was easy to combine with lysozyme [Chen et al., 2005]. CMC films showed mild inflammatory reactions, fast *in vivo* degradation and complete disappearance of inflammatory reactions in the end. These results demonstrated good biocompatibility of CMC. The carboxyl groups introduced in CM-chitosan destroy their sub-structure, reduce the crystal degree and increase the solubility. Moreover, CMC exhibited polymeric property of zwitterions. Hence, they have different dimensional structures and configurations, which endow them with several biologic activities [Chen et al., 2000]. CMC at low concentrations revealed stimulative effect on fibroblasts' growth, and exhibited good biodegradability and biocompatibility [Chang et al., 2008].

CMC with DS ranging from 0.25 to 1.19 was synthesized by alkalization of chitosan, followed by carboxymethylation with monochloroacetic acid. CMC alone could be electrospun into fibers and required the addition of a water-soluble polymer. Cross-linking by heat-induced esterification (at 140° C for 30 min) rendered CMC/PVA fibrous membranes insoluble in water. The mass retention and fiber morphology confirmed their highly substituted nature, which favored inter-molecular crosslinking leading to a more stable and water-insoluble fibrous membrane. However, the membrane from the less substituted CMC (DS = 0.36) was more hydrophilic and retained the desirable amine functionality, which is recognized to be responsible for antibacterial properties and biocompatibility [Du and Hsieh, 2008].

Guangyuan et al. [2009] used a combination of carboxymethylation and a bimodal molecular weight distribution for synthesizing chitosan derivatives. Specifically, chitosan was carboxymethylated to a theoretical extent of ~30% at physiological pH. CMC was used to form films and constructed by varying the ratio of high to low molecular weight (MW) fractions while maintaining the mechanical properties of the polymer. The rate of degradation of these films was found to be dependent upon both the carboxymethylation and the ratio of high to low MW polymer. Subsequently, biocompatibility was examined to determine the effects of the modifications upon Neuro-2a cells cultured on the films. Neuro-2a cells adhered and proliferated on the modified films at a rate comparable to those cultured on unmodified films. This data indicated that CMC exhibited tunable degradation rates and hence, could be used for neural tissue engineering.

With the development of biomaterial and biomedicine industry, CMC is expected to play an important role in both research and application fields.

Chapter 7

CYCLODEXTRIN LINKED CHITOSAN

Chitosans bearing cyclodextrin (CD) have been developed to form non-covalent inclusion complexes with a number of guest molecules altering their physicochemical properties for improved drug delivery, cosmetics, and analytical chemistry [Tanida et al., 1998; Prabaharan and Mano, 2006]. There are different means to link CD to chitosan e.g using 2-O-formylmethylated CD; inclusion complexes with 4-tert-butyl benzoic acid; chloroacyl CD in organic basic solvents or using 1,6- hexamethylene diisocyanate, 2-hydroxypropyl moiety, reducing sugar derivatives as spacer (Table 8). Chitosan derivative bearing β-CD moiety was also synthesized by the reaction between succinated chitosan and mono-amino- β-CD (Figure 2). The reaction was carried out in water at room temperature with water soluble carbodiimide. The obtained material was insoluble in water and contained 50% of cyclodextrin moiety [Aoki et al., 2001].

Chitosan microspheres obtained by cross-linking with glutaraldehyde were further reacted with chloroacyl CDs in organic basic solvents. Higher quantity of acyl CD could be linked to the microspheres through spacer and C–N bonds with a smaller cross-linking degree. The inclusion efficiency was checked with nalidixic acid, piroxicam, and p-nitrophenol [Mocanu et al., 2004]. The CD-linked chitosan can also be prepared by reaction with monochlorotriazinyl derivative of CD. Triazinyl moiety acts as a spacer [Martel et al., 2001]. This compound was used for decontamination of water containing textile dyes. El-Tahlawy et al. [2006] used a novel technique for preparation of β-CD grafted chitosan by reacting β-CD citrate with chitosan dissolved in formic acid solution and evaluated these polymers as antimicrobial agents. They also reported analogous synthesis with β-CD-itaconate and chitosan along with its utility as ion exchange resin [Gaffar et al., 2004]. The β-CD linked chitosan using 1,6-hexamethylene diisocyanate as spacer was also prepared [Sreenivasan, 1998; Chen et al., 2007]. This material interacted with cholesterol and was found to be useful as an adsorbent. The spacer can also be 2-hydroxypropyl moiety introduced by grafting β-CD onto chitosan using epoxyactivated chitosan [Zhang et al., 2004] or a reducing sugar derivative [Auzely-Velty and Rinaudo, 2002]. Aime et al. [2006] functionalized CD by means of a maleic spacer, whose free carboxyl group was subsequently activated with a carbodiimide to form amide linkages with amino groups of chitosan. The regioselectivity of the coupling could be accurately controlled if the 6-monotosyl-CD derivative was used as substrate for nucleophilic substitution with sodium maleate. An insoluble cross-linked chitosan bearing β-CD was prepared using N-succinyl chitosan and aminated-β-CD (mono-6-aminomono-6-deoxy-β-cyclodextrin) via amide bond formation in the presence of the water-soluble 1-ethyl-3-(3-dimethylaminopropyl) carbodiimide (EDC) under homogeneous conditions [Aoki et al., 2003].

The chitosan derivative bearing β-CD and Schiff-base was prepared by reacting chitosan with glutaric diketone, functionalizing with epichlorohydrin followed by reaction with β-CD. The synthesized polymer could adsorb metallic ions and phenolic compounds simultaneously [Zha et al., 2007].

Chitosan and β-CD/grafted chitosan, having different molecular weights, were evaluated as antimicrobial agents for different microorganisms such as, *Bacillus megaterium, Pseudomonas fragi, Bacillus cereus, Staphylococcus aureus, Escherichia ecoli* and *Aeromonas hydra* [El-Tahlawy et al., 2006]. In order to develop a treatment method for industrial wastewater, the adsorption of 4-nonylphenol ethoxylates (NPEs), non-ionic surfactants used in the industry onto chitosan beads having cyclodextrin (CDC beads) was investigated. The CDC-beads with CD content above 30% w/w were prepared by the condensation of carboxymethylated α, β, or γ -CD and chitosan beads. Among the different sizes of the CD cavities, β–CD was the best for the adsorption of NPE. The adsorption of NPE depended on the ethoxy chain in the NPE molecules. The used β -CDC beads could be recovered by ethanol treatment. The NPE could be continuously removed from a water solution using the β -CDC beads packed in a glass column [Aoki et al., 2007].

Table 8. Preparation, properties and applications of cyclodextrin linked chitosan

Derivative/Modification	Reaction with	Example	Property	Application	References
Cyclodextrin linked chitosan	2-O-formyl-methyl-α-CD by reductive N-alkylation	α-CD linked chitosan			Tanida et al., 1998; Tojimaa et al., 1999
	Reductive amination with 4-tert-butylbenzoic acid	Chitosan bearing pendant chitosan			Auzely-Velty and Rinaudo, 2001
		Supermolecular assemblies of chitosan			Auzely-Velty and Rinaudo, 2002
	Mono chloro triazinyl derivative of CD	CD linked chitosan		For decontamination of water containing textile dyes	Martel et al., 2001
	Tosylated β-CD	CD-linked chitosan	Enhanced solubility	Improved *in vivo* release	Chen and Wang, 2001
	Chloroacyl CD in organic basic solvents		Higher amounts of acyl CD are linked to the microspheres through spacer	Improved Inclusion efficiency of Nalidixic acid, piroxicam, p-nitro phenol	Mocanu et al., 2004
	β-CD itaconate			Ion exchange resin	Gaffar et al., 2004
CD itaconate-*graft*-chitosan	1,6- hexamethylene diisocyanate as spacer	β-CD linked chitosan	Interacts with cholesterol	Adsorbent	Sreenivasan, 1998; Chen et al., 2007
	2-hydroxypropyl moiety as spacer				Zhang et al., 2007

Table 8. (Continued).

Derivative/Modification	Reaction with	Example	Property	Application	References
	Reducing sugar derivative as spacer		Regioselectivity of coupling could be controlled by using 6-monotosyl-CD derivative as spacer		Auzely-Velty and Rinaudo, 2002
	N-succinyl chitosan and amidated β-CD via amide bond formation	Cross linked chitosan bearing β-CD			Aoki et al., 2003
	β- and γ-CDs are linked to chitosan through succinyl bridges		Bitter taste masked		Binello et al., 2004
	Chitosan microspheres, obtained through crosslinking with glutaraldehyde of an acetic acid solution of chitosan, in an organic suspension medium, were reacted with chloroacyl cyclodextrins in organic basic solvents	Chitosan derivatives containing cyclodextrin moieties as pendant groups		Conjugates used as supports for chromatographic separations or controlled release drug systems	Georgeta et al., 2004

Derivative/Modification	Reaction with	Example	Property	Application	References
	Reaction of chitosan microspheres and mono-(6-p-tosyl)-β-CD	Chitosan microspheres grafted with β-CD	Grafting and inclusion was found to be stable		Wang et al., 2005
	Coupling β-CD citrate with chitosan dissolved in different formic acid solutions having different concentrations	β-CD-grafted chitosan		Antimicrobial activity for microorganisms such as, *Bacillus megaterium, Pseudomonas fragi, Bacillus cereus, Staphylococcus aureus, Escherichia ecoli* and *Aeromonas hydra*.	El-Tahlawy et al., 2006
	SOCl₂ and CD were heated with chitosan (in DMF) and product was dried with 5% acetic acid.	Carboxymethyl-β-CD grafted chitosan	Introduction of chitosan enhanced the adsorption ability and adsorption selectivity of β-CD for guanosine 5′-monophosphate.	Application in separation, concentration and analysis of nucleotides in biological sample	Xiao et al., 2006
	CD-monoaldehyde	CD-chitosan derivative		Mucoadhesive	Venter et al., 2006
	β-CD citrate with chitosan dissolved in formic acid solution	β-CD grafted chitosan		Antimicrobial agent	El-Tahlawy et al., 2006

Zhang et al. [2009] investigated the possibility of using chitosan bearing β-CD nanocomplexes for controlled protein release. Synthesis was done by reacting N-succinylated chitosan with mono (6-(2-aminoethyl) amino-6-deoxy)-β-cyclodextrin in the presence of the water-soluble carbodiimide. The amount of β-CD grafted was up to 62.1% w/w. *In vitro* cytotoxicity study against NIH 3T3 cells showed that the complex was not cytotoxic. Self-assembled nanocomplexes containing insulin ranging from 190 nm to 328 nm, electrical charge from +3.7 to +25.5 mV and high loading efficiency of 37.7% could be prepared. Insulin release *in vitro* was affected by the medium pH and the composition of copolymer. The results demonstrated that complexed copolymer was a new promising vehicle for controlled protein release.

Nanoparticles made of chitosan and carboxymethyl-β-CD (CM-β–CD) were evaluated for their potential for delivery of macromolecular drugs. Chitosan and CM-β-CD or mixtures of CM-β-CD/tripolyphosphate (TPP) were processed to nanoparticles via the ionotropic gelation technique. The resulting nanoparticles ranged from 231 nm to 383 nm and showed a positive zeta potential ranging from +20.6 to +39.7mV. These nanoparticles were stable in simulated intestinal fluid pH 6.8 at 37° C for at least 4 h. Insulin and heparin (macromolecular model drugs) were incorporated into the different nanocarriers with association efficiencies of 85.5–93.3 and 69.3–70.6%, respectively. The association of these compounds led to an increase of the size of the nanoparticles (366–613 nm), with no significant modification of their zeta potentials (+23.3 to +37.1 mV). The release profiles of the associated macromolecules were highly dependent on the type of molecule and its interaction with the nanomatrix. Insulin release was observed to be high (84-97% released within 15 min) whereas, heparin remained highly associated to the nanoparticles for several hours (8.3–9.1% released within 8 h). Hence, these nanoparticles were suggested to be suitable for the fast or slow delivery of macromolecules [Krauland and Alonso, 2007].

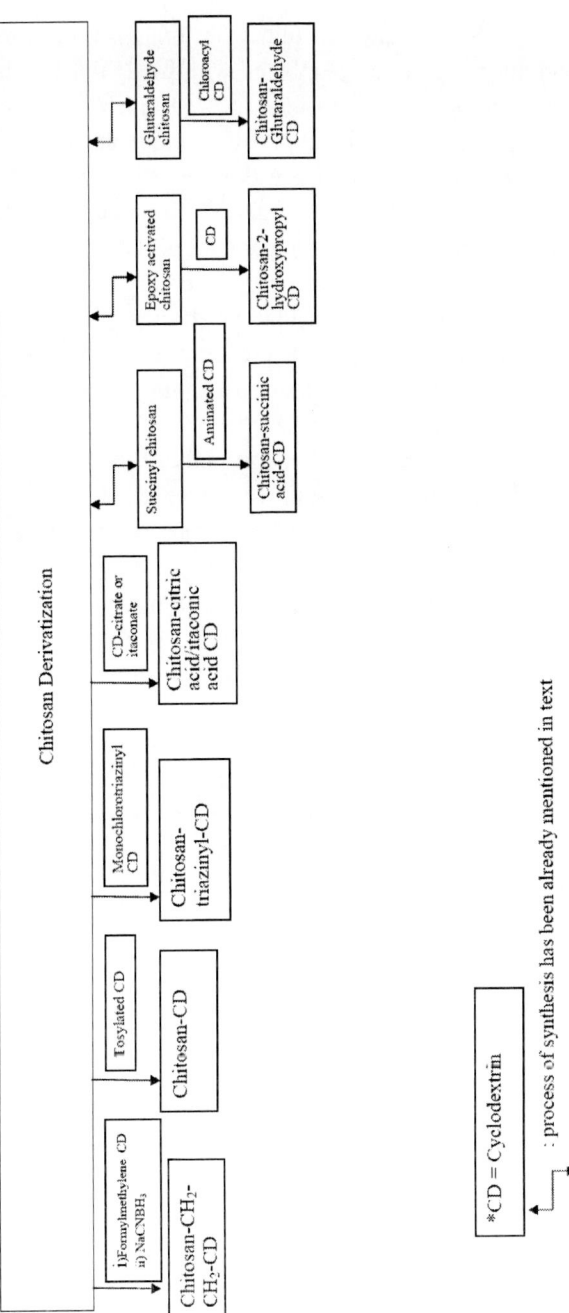

Figure 2. Approaches for modifying chitosan.

Moses et al. [2000] attempted to complex insulin with β-CD followed by encapsulation in the chitosan matrix. Insulin complexed with β-CD for 20 min was used for encapsulation in chitosan. The results showed that the matrix yielded different drug release profiles in simulated intestinal medium (pH 7.4). The change in the loading character of the matrix was found to be inversely related to the concentration of β-CD when it was above the stoichiometric equivalent of the drug. In an attempt to increase the payload of the drug in the matrix, the pH of the processing medium consisting of calcium chloride and chitosan was varied. It was found that the encapsulation efficiency increased with decrease in pH from pH 6 to pH 4. Loading efficiency was also increased by reducing the concentration gradient between the crosslinking medium and the alginate solution containing the drug.

Another study revealed that the supramolecular assembly of the polymers and α-CD molecules led to hydrogel formation in aqueous media. The poly(ethylene glycol) side chains on the chitosan backbones were found to form inclusion complexes with α-CD, creating hydrophobic micro-domains with channel-type crystalline structure, which play an important role as physical junctions in the hydrogels [Huh et al., 2004]. It was reported that these supramolecular hydrogels could be useful for biomedical applications because of their biocompatible constituents and supramolecular functionality, such as a thermo-reversible gel–sol transition property.

Chitosan-CD hybrid nanoparticles were obtained by the ionic gelation process in the presence of glutathione, chosen as a model drug. The nanoparticles were proven to efficiently encapsulate glutathione in their inner cores, thus showing promising perspectives as drug carriers [Ieva et al., 2009].

Chapter 8

N-ACYL CHITOSAN

N-Acyl derivatives of chitosan can be easily obtained from acyl chlorides and anhydrides (Figure 3). In a general way, acylation reactions occur easily in aqueous acetic acid/methanol, pyridine, pyridine/chloroform, trichloroacetic acid/dichloroethane, ethanol/methanol mixture, methanol/formamide or Dimethylacetamide (DMA) (containing 5-9% LiCl i.e. DMAc/LiCl) solvent systems (Table 9) [Shigemasa et al., 1999]. Due to fairly different reactivities of the two hydroxyl and the amino group on the repeating unit of chitosan, acylation can be controlled at the expected sites, i.e. on either amino [Seo et al., 1989; Hirano et al., 2002; Tien et al., 2003], hydroxyls [Sashiwa et al., 2002], or on both groups [Grant et al., 1990; Zong et al., 2000; Seo et al., 2001; Wu et al., 2005]. The introduction of hydrophobic branches generally endows new physicochemical properties such as the formation of polymeric assemblies, including gels [Martin et al., 2002], polymeric vesicles [Wang et al., 2001], Langmuir–Blodgett films [Nishimura et al., 1993; Xu et al., 1996], liquid crystals [Rout et al., 1993; Wu et al., 2003], membranes [Seo et al., 1995], and fibers [Hirano et al., 2000; Hirano and Moriyasu, 2004]. Hydrophobic associating water-soluble polymers have emerged as a new class of industrially important macromolecules. Some of these are intended to mimic the endotoxins [Desbrieres et al., 1996]. The introduction of hydrophobic branches also endows the polymers with a better soluble range than chitosan itself. Zong et al. [2000] synthesized acylchitosan with longer chains by reacting chitosan in pyridine/chloroform with hexanoyl, decanoyl, and lauryl chlorides. These acylated chitosans with 4 degree of substitution per monosaccharide ring (disubstitution at amino and monosubtitution each at hydroxyl groups) exhibited an excellent solubility in organic solvents such as chloroform, benzene, pyridine, and THF. The analysis indicated that these polymers form a layered structure in solid state and the layer spacing increases linearly with increasing length of side chains.

The presence of such layered structure was elucidated with N-aliphatic acyl chitosans and N-aliphatic-O-dicinnamoyl-chitosans with acyl as acetyl, butyryl, octanoyl, lauroyl and stearoyl moieties. Interestingly, none of these polymers could be dissolved in $CHCl_3$, CH_2Cl_2, THF, $(Me)_2CO$, DMAc, DMF, DMSO, DMSO/ $CHCl_3$. The reason for the striking stability of N-aliphatic acyl against solvents was attributed to the compact arrangement of both the main chains and the side chains of N-aliphatic acyl to form a crystal with strong hydrogen bond interactions together with strong interactions between closely packed hydrophobic side chains. On the other hand, the polymers belonging to the series of N-aliphatic- O-dicinnamoyl-chitosans displayed solubilities strongly related to the length of the flexible side chains. In general, increasing length of the flexible side chains reduced the solubility [Wu et al., 2006]. N-acylated chitosans with saturated (e.g. C2–C18) and unsaturated acyl groups of different chain length (e.g. oleic, linoleic, elaidoic, erucoyl) as well as aromatic acyl groups (e.g. phthaloyl, p-nitrobenzoyl, cinnamoyl) had been successfully synthesized to obtain randomly distributed substituents along the chitosan chain [Hirano and Ohe, 1975; Hirano et al., 1976; Hirano and Moriyasu, 1981].

Cyclic acid anhydrides too are used for acylation purpose via ring-opening reactions giving N-carboxyacyl chitosans (e.g. succinic, maleic, glutaric, itaconic, phthalic, cis-1,2,3,6-tetrahydrophthalic, 5-norbornyl-endo-2,3-dicarboxylic, cis-1,2-cyclohexyl dicarboxylic, trimellitic anhydride, (2-octen-1-yl)succinic, citraconic, trimellitic, pyromellitic) [Sashiwa and Shigemasa, 1999; Hirano and Moriyasu, 2004]. Thermolysis has been used for synthesizing acylated chitosan derivatives [Toffey and Glasser, 2001]. This method was used to prepare chitosan amides derived from acids, such as acetic, acrylic, methacrylic, trifluoroacetic, and myristic [Vasnev et al., 2006]. N-acylation of chitosan with longer chain acid chlorides increased its hydrophobic character and made important changes in its structural features, which were reflected in improved mechanical properties of tablets prepared using these derivatives. The release characteristics of the drug suggested that release was controlled by diffusion or by swelling followed by diffusion, depending on both the acyl chain length and the degree of acylation [Tien et al., 2003].

Table 9. Preparation, properties and applications of acyl derivatives of chitosan

Derivative/Modification	Reaction with	Example	Property	Application	References
N-acyl chitosan Acyl chitosan	Acyl chlorides and anhydrides (acylation)		These polymers form a layered structure in solid state and layer spacing increases with increasing length of side chains. Better solubility	Targeted drug delivery to cancerous cells (acylation with folic acid)	Zong et al., 2000; Lee et al., 2006
	Thermolysis of acyl ammonium salts in solid state		Improved mechanical properties of tablets. The release was found to be controlled by diffusion or by swelling followed by diffusion, depending on acyl chain length and degree of acylation		Toffey and Glasser, 2001; Tien et al., 2003
Hydroxyacyl chitosan	Carboxy methylation	Hexanoyl chitosan	Excellent water absorption and water retention abilities under neutral conditions and used as a carrier for delivering amphiphatic agents.		Liu et al., 2006

Table 9. Continued

Derivative/Modification	Reaction with	Example	Property	Application	References
Carboxyacyl chitosan	Chitosan with hexanoyl chloride in a mixture of pyridine and THF	Hexanoyl chitosan			Winie et al., 2004.
	Acrylic acid and NaOH	*N*-carboxyethyl chitosan			Weng et al., 2008
		Formyl, acetyl, propionyl, butyryl, hexanoyl, acetanoyl, decanoyl, dodecanoyl, tetradecanoyl, lauroyl, myristoyl, palmitoyl, stearoyl, benzoyl, monochloroacetyl, dichloroacetyl, trifluoroacetyl, carbamoyl, succinyl, acetoxybenzoyl		Textiles, membranes, and medical aids	Gupta and Kumar, 2000

N,N-Diacyl Chitosan

N,N-Diacyl Chitosan	Pyridine/chloroform with carboxylic acid chlorides; and acylchitosan has been prepared by chemical N-acylation of the chitosan with the corresponding fatty acid anhydride. Acylation reactions are carried out frequently in aqueous acetic acid/methanol	N-acyl chitosan	Hydrophobic derivatives formation. Swelling index of chitosan derivatives decreasd with the increase of the degree of substitution hence, tighter and more compact structure for limiting water uptake	Expected to be good models of drug delivery systems.	Hirano and Ohe, 1975; Fujii et al.,1980; Hirano and Midorikawa 1998, Hirano et al., 2002; Rodrigues, 2005
	Pyridine and chloroform followed by hydrolysis	N,N-Diacyl Chitosan			Ming-chun et al., 2005
	Carboxy methylation	Hexanoyl chitosan	Excellent water absorption and water retention abilities under neutral conditions and used as a carrier for delivering amphiphatic agents.		Liu et al., 2006

Table 9. Continued

Derivative/Modification	Reaction with	Example	Property	Application	References
O-acyl chitosan R = alkyl chain O-(acyl) chitosan	Introduction of hydrophobic moiety with an ester linkage	O,O-didecanoyl-chitosan, O-succinyl chitosan	Hydrophobic groups contribute organo solubility The ester linkage is hydrolysed by enzyme like lipase. Therefore are biodegradable coating material		Sashiwa et al., 2002a; 2002b; Kim et al., 2003
	Alkanoic acid derivatives with chitosan in the presence of H2SO4 as a catalyst	O-(butyroyl) chitosan		Fungicidal activity against the grey mould *Botrytis cinerea* (Leotiales: clerotiniaceae) and the rice leaf blast pathogen *Pyricularia grisea*	Badawy et al., 2005

Hexanoyl chitosan with carboxymethylation was developed into amphiphatic hydrogel with excellent water-absorption and water retention abilities under neutral conditions and then employed as a carrier for delivering amphiphatic agents [Liu et al., 2006]. The hexanoyl substitution significantly improved the water-absorption ability of hydrogel by altering the number of water-binding sites under low humidity. In addition, the state of water in fully swollen state retarded the water mobility during deswelling, and enhanced amphiphatic drug encapsulation efficiency compared to pristine chitosan.

The acylated chitosan are being applied for stabilization of nanoparticles of iron oxide, and gold [Remantbahadur et al., 2006; Bhattarai et al., 2007]. N-succinyl-chitosan obtained by introduction of succinyl groups into N-terminal of the glucosamine units of chitosan could be modified easily with respect to succinylation degree by changing reaction conditions and the molecular weight using hydrochloric acid [Yamaguchi et al., 1981; Kato et al., 2002]. Although, N-succinyl-chitosan was initially developed as wound dressing material [Kuroyanagi et al., 1994], it is currently also used as a cosmetic material [Izume, 1998]. New wound dressings composed of N-succinyl-chitosan and gelatin have also been developed [Tajima et al., 2000]. N-succinyl-chitosan has unique *in vitro* and *in vivo* characteristics due to many carboxyl groups. For example, ordinary chitosan can be dissolved in acidic but not in alkaline water, whereas, N-succinyl-chitosan exhibits the opposite behaviour [Kato et al., 2002]. Further, it is valuable as a carrier as it can readily be used for conjugating with various drugs due to $-NH_2$ and $-COOH$ groups in its structure e.g. carbodiimide and mitomycin C [Song et al., 1992; Kato et al., 2004]. N-succinyl chitosan can form well-dispersed and stable nanospheres that exhibit great potential in the controlled drug delivery e.g oxymatrine [Zhu et al., 2006; Yan et al., 2006]. The succinyl chitosan can be adapted for solubility by addition of a long alkyl moiety as hydrophobic function to the amino group given that succinyl moiety provides hydrophilic and alkyl moiety provides the hydrophobic properties. This adapted derivative was used as a carrier for doxorubicin [Xu et al., 2007]. For liver-targeting lactose group was introduced [Kato et al., 2001a; 2001b]. N-acetyl, N-propionyl and N-hexanoyl with different degrees of substitution were synthesized and evaluated for *in vitro* antibacterial activity [Hu et al., 2007]. The nanoparticles prepared from N-acyl chitosan were found to be compatible with blood [Lee et al., 2004]. The betaine derivatives of chitosan have two major advantages over chitosan. They are water-soluble at physiological pH, and they have a permanent positive charge on the polysaccharide backbone. Condensation with carbodiimide of N-acylated chitosan was performed with amino acids (lysine, arginine, aspartic acid, phenylalanine) and these moieties were subsequently entrapped in PLA surfaces that demonstrated good cyto-compatibility to chondrocytes [Zhu et al., 2002]. Acylated chitosan when complexed with DNA revealed enhanced transfection efficiency due to enhanced cell membrane–carrier interaction and/or destabilization of cell membrane [Kim et al., 2001; Yoo et al., 2005] making it feasible to target the DNA delivery to cancerous cells [Mansouri et al., 2006; Lee et al., 2006].

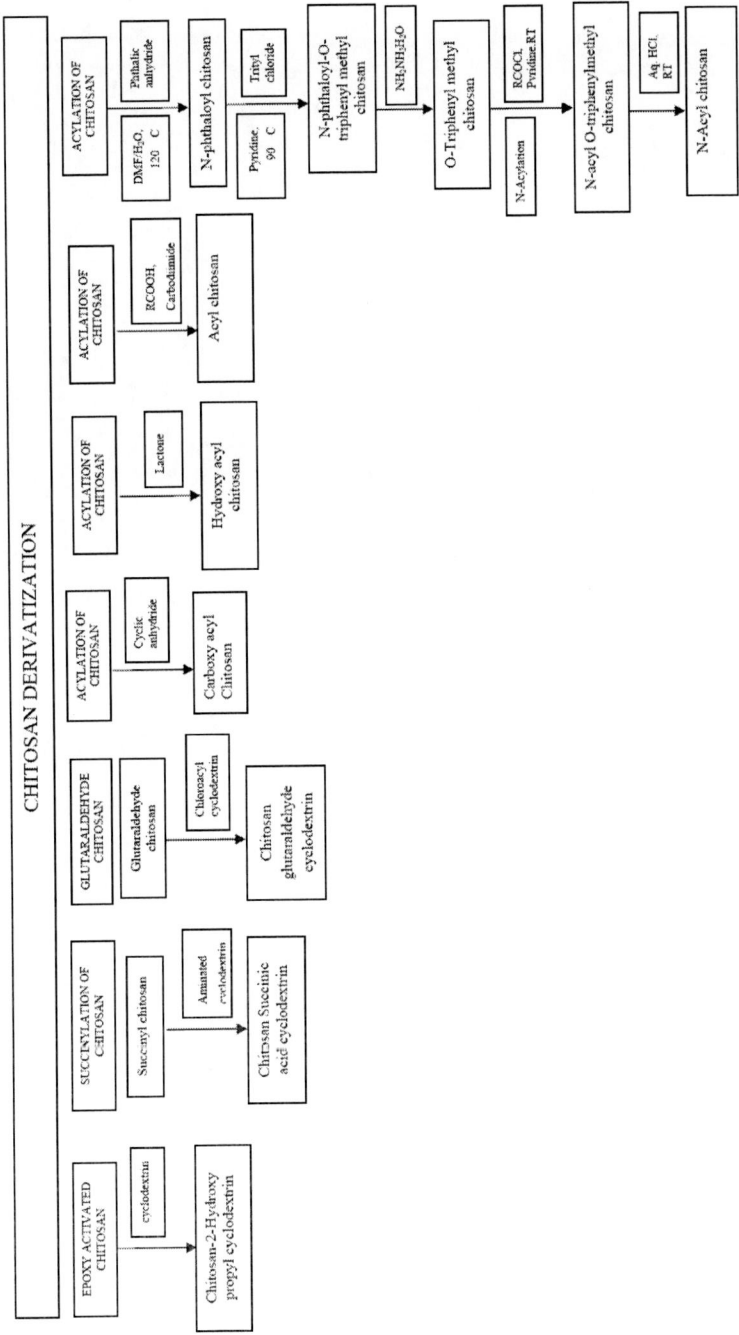

Figure 3. Approaches for modifying chitosan.

Chapter 9

GRAFT COPOLYMERIZATION
OF CHITOSAN

Graft copolymerization is well-known method for the modification of chitosan and represents convenient and effective way of improving the physical and mechanical properties for practical uses. Graft copolymerization is one of the important techniques used to modify chitin and chitosan chemically as well as for widening their therapeutic and other uses. Different reagents/initiators have been developed for graft copolymerization e.g. ammonium persulphate [Lv et al., 2009], potassium persulphate [Akgün et al., 2007], cerric ammonium nitrate [Joshi and Sinha, 2007], thiocarbonation potassium bromate [El-Tahlawy and Hudson, 2001], potassium diperiodatecuprate [Liu et al., 2002], bisisobutyronitrile [Deng et al., 2009] and ferrous ammonium sulphate [Lagos and Reyes, 2003]. In addition graft copolymerization can also be prepared via radical generation [Lv et al., 2009], polycondensation [Yang et al., 2005], oxidative coupling [Romaškevič et al., 2007], ring opening [Liu et al., 2005], grafting onto method [Huang et al., 2005] or photochemical reaction [David and Samuel, 2002]. The parameters like grafting percentage or the efficiency depends upon type and concentration of initiator, monomer, concentration, reaction temperature and time. Many studies have been conducted by different researchers to study the effects of the above mentioned variables on grafting parameters and on the grafted chitosan [Kim et al., 2000; Sun et al., 2003; Xie et al., 2002a].

GRAFT COPOLYMERIZATION BY FREE RADICALS

Various redox reagents have been reported for carrying out graft polymerization onto chitosan e.g. cerric ammonium nitrate, potassium persulphate, ammonium persulphate or Fenton's reagent. Several studies have been conducted on copolymerization after primary derivatization of chitosan e.g. carboxymethyl chitosan [Joshi and Sinha, 2007], N-carboxymethyl chitosan [Singh and Ray, 1994; Kang et al., 2006], maleolyl chitosan [Don and Chen, 2005], hydroxypropyl chitosan [Xie et al., 2002b], or trimethyl chitosan [Mao et al., 2005] etc. Chitosan was modified with polyacrylic acid, using a grafting reaction in a homogeneous phase [Yazdani-Pedram et al., 2000]. The grafting was carried out in the presence of potassium peroxydisulfate (PPS) and ferrous ammonium sulfate (FAS) as the combined redox initiator system. The efficiency of grafting was found to depend on monomer, initiator, and ferrous ion concentrations, as well as on the reaction time and temperature. It was observed that the level of grafting could be controlled to some extent by varying the amount of ferrous ion as a co-catalyst in the reaction. The maximum efficiency of grafting attained was 52%. The results revealed that in homogeneous systems the grafting reactions take place not only on the surface but also in the molecules of the whole substrate. A graft material based on chitin and poly acrylic acid has been prepared by using potassium peroxodisulfate as a redox initiator. The water sorption ability of the films revealed that the films possessed a network structure and exhibited enhanced water sorption ability [Tanodekaew et al., 2004].

GRAFTING VIA RADIATION

Grafting with the aid of radiation has been of great interest. Pengfei et al. [2001] used ^{60}Co-γ-radiation for grafting polystyrene onto chitosan at ambient temperature. The study revealed the effect of various variables such as adsorbed dose, type of solvent and effect of oxygen. Grafting yield was found to increase with increase in the adsorbed dose. Other researchers reported the use of N, N'-dimethyl amino ethyl methacrylate for radiation grafting [Kang et al., 2006], butyl acrylate using γ-radiation [Li et al., 2005; Huang et al., 2005], polyhydroxy ethyl methacrylate by UV light [Ng et al., 2001] and polyacrylonitrile by microwave radiation [Singh et al., 2004]. The grafting percentage was increased with the dose but decreased with the increase in chitosan concentration or reaction temperature [Yu et al., 2003]. The grafted chitosan films were found to exhibit more hydrophobicity and impact strength. Microwave irradiation using poly acrylonitrile revealed good grafting yield (70%) in 1.5 min [Singh et al., 2005].

GRAFTING BY POLYCONDENSATION

Lactic acid [Bhattarai et al., 2006] or 4-dimethyl aminopyridine has been used for condensation polymerization. Lactic acid has been used to prepare a pH sensitive hydrogel whereas 4-dimethyl amino pyridine was used as grafting polymer so as to improve the adhesion, compatibility or to support the profileration of human endothelial cells [Feng and Dong, 2007].

Nanoparticles of ~10 nm in diameter made with chitosan or lactic acid-grafted chitosan were developed for high drug loading and prolonged drug release. Encapsulation efficiency of 92% and release rate of 28% from chitosan nanoparticles over a 4-week period were demonstrated with bovine serum protein. To further increase drug encapsulation, prolong the drug release and increase its solubility at neutral pH, chitosan was modified with lactic acid by grafting D,L-lactic acid onto amino groups in chitosan without using a catalyst. The lactic acid-grafted chitosan nanoparticles demonstrated drug encapsulation efficiency of 96% and protein release rate of 15% over 4 weeks. With increased protein concentration, the drug encapsulation efficiency decreased and drug release rate increased. Unlike chitosan, which is generally soluble only in acid solution, the chitosan modified with lactic acid could be dissolved in solutions of neutral pH, offering an additional advantage of allowing proteins or drugs to be uniformly incorporated in the matrix structure with minimal degradation [Bhattarai et al., 2006].

GRAFTING BY EPOXY-TERMINATED POLYDIMETHYLSILOXANE

Epoxy-terminated polydimethylsiloxane (PDMS) was grafted onto chitosan using UV irradiation at room temperature without using a catalyst. The product was a pH-sensitive hydrogel without any chemical cross-linking. In fact, the PDMS substituents provided the basis for hydrophobic interactions that contributed to the formation of the hydrogel network. The hydrogels exhibited high equilibrium water content in the range of 82–92% [Kim et al., 2002].

Chitosan mini-emulsions were used for the synthesis of epoxy particles by polyaddition. For the preparation of mini-emulsions, chitosan composed of two different molecular weights (low and high) was dissolved in 1% v/v aqueous acetic acid, which was used as aqueous phase in the mini-emulsion polymerization. The monomer phase consisting of a mixture of styrene, hexadecane as hydrophobe, and 2,2-azobis-(2,4-dimethylvaleronitrile) as oil-soluble initiator was added to the chitosan solution. The mixture obtained was stirred at room temperature and mini-emulsification was obtained by ultrasonication. As chitosan bears amine and alcohol functions, it can react with the epoxide and can be grafted onto the particles, which are obtained by polyaddition reaction. This is a convenient technique to modify water-soluble chitosan with water insoluble reaction partners, thus resulting in new chitosan derivatives [Marie et al., 2002].

MISCELLANEOUS GRAFT COPOLYMERS OF CHITOSAN

The term PEGylation usually refers to a process involving the conjugation of PEG with a substrate. The conjugation of PEG to drugs, especially protein drugs, is well known to enhance the solubility and stability of the protein in solution, to alter bioavailability, pharmacokinetics, immunogenic properties, and biological activities, and also to protect the drugs from recognition by the immune system, prolonging circulation time and efficacy *in vivo* [Saito et al., 1997]. Several methods have been reported on the PEGylation of chitin/chitosan using PEGs with various terminal reactive groups [Harris et al., 1984]. Several other techniques are also reported for grafting of polymers e.g. the conductive polymers were prepared by grafting polyaniline onto chitosan by oxidative coupling. Copolymerization by the mechanism of ring opening of polymer was carried out by Jenkins and Hudson [2001] using N-carboxyanhydride. The polymerization of aniline, an air stable organic conductive material, in the presence of chitosan resulted in a graft copolymer that was soluble in a slightly acidic aqueous solution and formed self-supporting materials, including thick films and fibers, that were conductive when protonically doped. The reaction of chitosan in aqueous acetic acid solution with aniline in the presence of ammonium persulfate (APS) enables the introduction of polyaniline side chains at the amino groups. Chitosan-graft-polyaniline was fabricated into films and fibers, but the properties varied according to the ratio of the amino group to aniline in the grafting reaction. With the ratio of 1:1 to 1:5, the products were sturdy and flexible, while those with a ratio of 1:6 to 1:10 were brittle. Optical microscopic observations indicated that the products prepared at a ratio below 1:5 were homogeneous, but those above 1:6 had crystalline regions [Yang et al., 1989].

Chapter 10

BIOMEDICAL APPLICATIONS

CONTROLLED RELEASE DRUG DELIVERY SYSTEMS

Chitosan has interesting biopharmaceutical characteristics such as pH sensitivity, biocompatibility and low toxicity. Moreover, chitosan is metabolised by certain human enzymes, especially lysozyme, and is biodegradable. Due to these favorable properties, the interest in chitosan and its derivatives in drug delivery applications has increased in recent years. Kumbar and Amabhavi [2003] used the microspheres of polyacrylamide grafted chitosan cross-linked with glutaraldehyde to encapsulate indomethacin, a nonsteroidal anti-inflammatory drug, used in the treatment of arthritis. The microspheres with a mean particle size of 525 nm were produced by the water/oil emulsion technique, and encapsulation of indomethacin was carried out before cross-linking the matrix. The release of indomethacin depended on the cross linking of the network and also on the amount of drug loaded. Microspheres of grafted chitosan cross linked with glutaraldehyde were prepared to encapsulate nifidifine. *N*-laurylcarboxymethyl chitosan with both hydrophobic and hydrophilic groups was studied in connection with the delivery of taxol to cancerous tissues [Yoshioka et al., 1995]. Other examples are related to the production of polymeric vesicles for encapsulation of hydrophobic compounds like bleomycin [Miwa et al., 1998]. Chitosan has also been modified with polyacrylonitrile through ceric-initiated graft copolymerization [Pourjavadi et al., 2003]. The chitosan-graft-poly(acrylonitrile) product was then saponified using sodium hydroxide aqueous solution to prepare a novel super-adsorbent hydrogel, H-chitosan-polyacrylonitrile. It was observed that both modified chitosans exhibited enhanced thermal stability over chitosan. Swelling capacity of the hydrogel was observed to be affected by NaOH concentration. The lower concentration of alkali resulted in higher water absorbency. The super-absorbent swelling gel exhibited a high sensitivity to pH. Sharp and large volume changes were observed with small pH variation. This super-adsorbent polyampholytic network intellingently responding to pH may be considered as an excellent candidate to design novel drug delivery systems. The chitosan-grafted poly (vinyl alcohol) (PVA) copolymer matrix containing prednisolone exhibited sustained release charactoristics [Kweon and Kang, 1999]. In particular, systems that employed combination of chitosan and poly(N-isopropylacrylamide) showed drug release profiles that could be controlled by both pH and temperature [Lorenzo et al., 2005; Bhattarai et al., 2005]. Chitosan-based systems bearing β-CD cavities have been proposed as a matrix for controlled release [Prabaharan and Mano, 2005; Krauland and Alonso, 2007]. Due to the presence of the hydrophobic β-CD rings, the systems exhibited slower release of the entrapped hydrophobic drug. Stimuli-responsive hydrogels with improved drug loading capacity were observed to exhibit sustained release behaviour [Prabaharan and Mano, 2006].

TISSUE ENGINEERING

The recent tissue engineering research is based on the seeding of cells onto porous biodegradable polymer matrixes. A primary factor is the availability of good biomaterials to serve as the temporary matrix. Recently, chitosan and its derivatives have been reported as attractive candidates for scaffolding materials because they degrade as the new tissues are formed, eventually without inflammatory reactions or toxic degradation. Possible applications of chitosan in spine tissue engineering encompass three different fields, namely spine fusion, gene therapy, and nucleus pulpous regeneration. When a bone graft alternative is applied during spinal fusion procedures, several local biomechanical factors are considered, depending on the type and position of the chosen graft. Anterior inter-body grafts are exposed to high compressive forces and need to possess load-bearing ability. In contrast, a posterior-applied bone graft is placed along the tension side of the spinal column in the absence of local compressive stimuli, and thus bone graft incorporation is less likely to be affected by local biomechanical factors [Mwale et al., 2005]. Materials such as PLA (poly-d,l-lactic acid) or PLA-PEG (polyethylene glycol) have been tested for spinal fusion and are considered good candidates due to their plasticity, stiffness, biodegradability, and ability to support cells and growth factor [Saito et al., 1999; Hoshino et al., 2007]. A possible application of chitosan could be a composite graft material with a predictable degradation rate and macroporous structure that would allow linking of growth factors and support osteogenic cells for spinal fusion [Prabaharan et al., 2006]. Intervertebral disks possess different functional and anatomic regions: the inner nucleus pulpous, a hydrated gelatinous tissue rich in proteoglycans, and the outer annulus fibrosis made of concentric collagen lamellae. Loss of proteoglycans and water content in the inner nucleus pulpous initiates degenerative spinal disease. Biologic disk regeneration is considered a promising approach to restore biological integrity and function of a pathologically altered disk [Boden, 2002]. Another approach regarding the chemical modification of chitosan for tissue engineering applications has been to introduce the specific recognition of cells by sugars. The synthesis of sugar-bound chitosan can be related to preparation of mannosylated chitosan having specific recognition to antigen presenting cells such as β-cells, dendritic cells and macrophages [Kim et al., 2006].

WOUND HEALING AND ANTIMICROBIAL PROPERTIES

Grafted chitosan presents interesting properties for wound-healing applications, because chitosan derivatives can exhibit enhanced bacteriostatic activity with respect to pure chitosan. Ethylene diamine tetra acetic acid (EDTA) grafted on chitosan increases the antibacterial activity of chitosan by chelating magnesium that under normal circumstances stabilizes the outer membrane of gram-negative bacteria [Valenta et al., 1998]. The increase in chitosan antimicrobial activity is also observed with carboxymethyl-chitosan, which makes essential transition metal ions unavailable for bacteria [Muzzarelli, 1989] or binds to the negatively charged bacterial surface to disturb the cell membrane [Liu et al., 2001]. Carboxymethyl-chitosan is used for reducing periodontal pockets in dentistry [Bernkop-Schnu"rch et al., 1997] and chitosan grafted with EDTA as a constituent of hydro-alcoholic gels for topical use [Muzzarelli, 1989]. Poly(ethylene terephthalate) (PET) fiber was treated with ^{60}Co-c-ray and grafted with acrylic acid. The resulting fibers were further grafted with chitosan via esterification. Collagen was then immobilized on chitosan grafted fibers. The antibacterial activity of chitosan against *Staphylococus aureus, Escherichia coli, and Pseudomonas aeruginosa* was preserved after collagen immobilization. The results indicated that by grafting with chitosan and immobilizing with collagen, PET fibers exhibited both antibacterial activity against four pathological bacteria and improved the proliferation of fibroblast [Jou et al., 2007]. Similar work done by Huh et al. [2001] reported that PET treatment with oxygen plasma glow discharge, followed by a graft copolymerization of acrylic acid could be used to prepare carboxylic acid group–containing PET. Chitosan and quaternized chitosan were then ionically or covalently immobilized on PET–acrylic acid. The experiments of antibacterial activity of chitosan graft-PET against *Staphylococus aureus* showed high growth inhibition in the range of 75–86% and still maintaind 48–58% bacterial growth inhibition after laundering. Hu et al. [2003] reported the grafting of acrylic acid on ozone-treated poly(3-hydroxybutyric acid) and poly(3-hydroxybutyric acid-co-3-hydroxyvaleric acid) membranes. The resulting membranes were further grafted with chitosan via esterification. These chitosan-grafted membranes showed antibacterial activity against *Escherichia coli, Pseudomonas aeruginosa*, methicillin-resistant *Staphylococcus aureus* (MRSA), and *Staphylococcus aureus*. Acrylic acid grafting increased the biodegradability with *Alcaligens faecalis*, whereas chitosan and chitooligosaccharide grafting reduced the biodegradability. In addition, chitosan-grafted-poly (3- hydroxybutyric-acid-co-3-hydroxyvaleric acid) membrane showed higher antibacterial activity and lower biodegradability than chitooligosaccharide-grafted membrane. Two anionic soluble monomers, mono(2-methacryloyloxyethyl) acid phosphate and vinyl sulfonic acid sodium salt can be grafted on chitosan to obtain copolymers with zwitterionic property [Jung et al., 1999]. The grafting reaction improved the antimicrobial activities of chitosan against *Candida albicans, Trichophyton rubrum,* and *Trichophyton violaceum*, which depended on the amount and type of grafted chains, as well as on the changes in pH [Yang et al., 1989]. The antibacterial activity of the polypropylene was enhanced by the modification of radiation–induced grafting of acrylic acid and the immobilization of chitosan onto the polypropylene- graft-acrylic acid-modified polymer [Yang and Lin, 2004].

Chitosan possess the characteristics favorable for promoting rapid dermal regeneration and accelerated wound healing. It was observed that chitosan oligosaccharides possessed a stimulatory effect on macrophages, and both chitosan and chitin were chemo attractants for neutrophils *in vitro* and *in vivo.* Chitin and chitosan may further facilitate wound healing by stimulating granulation tissue formation or re-epithelization [Campos et al., 2006]. The argon plasma–treated chitosan membranes exhibited excellent attachment, migration, and proliferation of the human skin–derived fibroblasts compared to the untreated ones. To improve the healing process, chitosan has been combined with a variety of modified materials such as growth factors, extracellular matrix components, and antibacterial agents. The incorporation of basic fibroblast growth factor with chitosan accelerated the rate of healing [Mizuno et al., 2003; Liu et al., 2007b]. Hence, it has been revealed that chitosan derivatives have great potential to be used in other biomedical applications.

MISCELLANEOUS APPLICATIONS

Chitosan-grafted systems have also been investigated for use in the cardiovascular science. Mao et al. [2004b] studied the blood compatible properties of O-butyrylchitosan-grafted poly(vinyl chloride) and polyethylene. The grafted polymers was observed to be compatible with blood [Mao et al., 2004a; 2004b]. The amphiphillic polymers poly(ethylene glycol) (PEG) or poly(ethylene oxide) (PEO) have been used with natural and synthetic macromolecules to obtain hydrophilic biocompatible materials. The surface modification of natural and synthetic macromolecules with water-soluble polymers such as PEG or PEO can prevent plasma protein adsorption, platelet adhesion, and thrombus formation by stearic repulsion mechanism [Ikada, 1984]. The chitosan membrane grafted with hydroxyethyl methacrylate by plasma polymerization in water did not exhibit significantly different bulk properties as compared to modified membrane, although the surface hydrophilicity was improved. It was found that the permeability of chitosan membrane could be controlled through plasma treatment, having the potential to be used in the dialysis process [Li et al., 2003]. Colon-targeted delivery of bioactives has recently gained importance in addressing specific needs in the therapy of colon-based diseases. Many approaches have been investigated for developing colon-specific delivery systems without much success. Recent research into the utilization of the metabolic activity and the colonic flora the lower gastrointestinal tract has attained great value in the design of novel colon-targeted delivery systems based on natural biodegradable polymers. Chitosan is a promising polymer for colon drug delivery since it can be biodegraded by the colonic bacterial flora, and it has muco-adhesive character

[Kaur et al., 2009a; 2009b]. Table 10 summarizes the different formulations prepared using various chitosan derivatives.

Table 10. A summary of different formulations prepared by employing chitosan derivatives

Derivative	System/formulation	Drug	Reference
Chitosan cross-linked with Dialdehyde, Glutardehyde	Transdermal delivery	Propanolol HC	Thacharodi and Rao, 1995
	Membranes for controlled delivery	Riboflavin	Aly, 1998
	Transdermal delivery	Oxprenolol HCl	Bolgul et al., 1998
	Fibers	5-fluorouracil	Denkbas et al.,2000
	Freeze-dried microspheres	Goserelin	Illum et al., 2000
Chemical crosslinking of the chitosan in the presence of glutaraldehydes	Nanoparticles	5-fluorouracil	Ohya et al., 1994
N-trimethylene chloride Chitosan		Octreotide	Thanou et al., 2000a
Chitosan succinate and Chitosan phthalate	Matrices for the colon-specific oral delivery	Sodium diclofenac	Sinha and Kumria, 2001
Chitosan cross-linked with Genipin	Injectable microspheres		Mi et al., 2002b
	Gel beads for the controlled release	Indomethacin	Mi et al., 2002a
chitosan glutamate	Nanoparticles	Insulin	Dyer et al., 2002
Chitosan cross-linked with glutaraldehyde, together with sodium alginate with or without microcrystalline cellulose	Mucoadhesive vaginal tablets	Metronidazole	Kamel et al., 2002
Polyacrylamide grafted chitosan crosslinked with glutaraldehyde	Microspheres	Indomethacin	Kumbar and Amabhavi, 2003
N-octyl-O-sulfate chitosan	Micelle	Taxol	Zhang et al., 2003
Chitosan cross-linked with glutaraldehyde	Microspheres with mucoadhesive property	Glipizide	Patel et al., 2005
Ionically cross-linking N-[(2-hydroxy-3-trimethylammonium) propyl] chitosan chloride in the presence of tripolyphosphate	Microgels	Methotrexate	Zhang et al., 2006
N-palmitoyl Chitosan	Micelle	Ibuprofen	Jiang et al., 2006

Table 10. Continued

Derivative	System/formulation	Drug	Reference
Hydroxylpropylcyclodextrins (ionic cross linking of chitosan with sodium tripolyphosphate)	Nanoparticles for transmucosal delivery	Triclosan and furosemide	Maestrelli et al., 2006
Chitosan and randomly methylated b-cyclodextrin	Nasal administration	Estradiol	Leng et al., 2006
50% N-acetylated low molecular weight chitosan (LMWC)	Renal targeted intravenous injection	Prednisolone	Yuan et al., 2007
N-trimethyl chitosan	Nanoparticles	Cisplatin	Cafaggi et al., 2007
Chitosan–poly (acrylic acid)	Hollow nanosphere	Doxorubicin	Hu et al., 2007a
Succinyl chitosan	Eudragit coated microspheres	Prednisolone	Oosegi et al., 2008
Chitosan-g-poly(N-isopropylacrylamideco-N,N-dimethylacrylamide)	Magnetite nanoparticles	Doxorubicin	Yuan et al., 2008
Quaternized derivatives of chitosan [trimethyl chitosan (TMC), dimethylethyl chitosan (DMEC), diethylmethyl chitosan (DEMC) and triethyl chitosan (TEC)]	Nanoparticles	Insulin loaded	Sadeghi et al., 2008
Methoxy poly (ethylene glycol) (PEG)-graftchitosan and lactose-conjugated PEG-graft-chitosan	Polyion complex micelles for liver-targeted Delivery	Diammonium glycyrrhizinate	Yang et al., 2009a
Chitosan-modified poly(lactic-coglycolic acid)	Nanoparticles	Paclitaxel	Yang et al., 2009b
Carboxymethyl chitosan modified with linoleic acid	Nanoparticles	Adriamycin	Tan and Liu, 2009
PEGylated chitosan	Nanoparticles	Paclitaxel, Camptothecin, Methotrexate, and trans- retinoic acid	Qu et al., 2009b; Opanasopit et al., 2006; Yang et al., 2008; Jeong et al., 2006
N-octyl-chitosan and N-octyl-O-sulfate chitosan	Tablets	Calcein, FITC-dextran	Green et al., 2009

Chapter 11

SAFETY AND TOXICOLOGICAL STUDIES

The regulatory and toxicological status of chitosan has already been established. Chitosan is widely regarded as a non-toxic, biologically compatible polymer [Thanou et al., 2001a]. It is approved for dietary applications in Japan, Italy and Finland [Illum, 1998] and it has been approved by the FDA for use in wound dressings [Wedmore et al., 2006]. The modifications made in chitosan could make it more or less toxic and any residual reactants should be carefully removed. It is important to consider that the formulation of chitosan with a drug may alter the pharmacokinetic and biodistribution profiles. For instance, in the case of chitosan/plasmid DNA nanoparticles, the *in vivo* kinetics and distribution are mainly controlled by the nanoparticle properties (size and charge). Further, cellular uptake kinetics may be altered due to the charge interaction (e.g. in the case of DNA complexes). This balancing, or reduction, of the positive charges on the chitosan molecule influences its interaction with cells and the microenvironment, often leading to decreased uptake and a decrease in toxicity. Schipper et al. described the effect of chitosans with differing molecular weights and degree of deacetylation [DA] on CaCo-2 cells, HT29-H and *in situ* rat jejunum. Toxicity was found to be dependent on DA and molecular weight. At high DA the toxicity was related to the molecular weight and the concentration, while at lower DA, toxicity was less pronounced and less related to the molecular weight. However, most of the chitosans tested did not increase dehydrogenase activity significantly in the concentration range tested (1–500 µg/ml) on Caco-2 cells. The *in situ* rat jejunum study showed no increase in LDH activity with any of the chitosans tested (50 µg/ml) [Schipper et al., 1996; 1999]. Trimethyl chitosan has relatively low cytotoxicity. It is an oligomer (Mw 3–6 kDa) with

IC 50 >10 mg/ml for degrees of quaternisation below 55%. However, it was shown that trimethyl chitosan of increasing degree of trimethylation possessed increased cytotoxicity as did higher molecular weight derivatives (100 kDa) [Kean et al., 2005]. Chitosan and its derivatives seem to be toxic to several bacteria [Jumaa et al., 2002], fungi [Guo et al., 2006] and parasites [Pujals et al., 2008]. This pathogen related toxicity is an effect which could aid in infectious disease control. When emulsions containing chitosans were tested, bacterial inhibition took place in acidic solutions (pH 5–5.3), and the 87 kDa having 92% degree of deacetylation (DD) chitosan was more effective than a 532 kDa having 73% DD chitosan against both *Pseudomonas aeruginosa* and *Staphylococcus aureus. In vivo* toxicity particularly after long term administration is of high importance for the design of drug delivery systems based on chitosan. Succinimidyl chitosan was not significantly toxic at 2 g/kg in mice [Song et al., 1993]. A photo cross linkable chitosan was developed as tissue glue. No toxic effect was noted after subcutaneous implantation of 200 μl of 30 mg/ml photo-cross linked azide–chitosan–lactose gel. This photo cross linkable gel exhibited the same or greater sealing capabilities in comparison to fibrin glue [Ono et al., 2000]. Wound contraction and faster healing were observed using both chitosan [Azad et al., 2004] and photo cross linked chitosan [Ono et al., 2000]. Slight toxicity at high dose was observed after oral administration of trimethyl chitosan/pDNA nanoparticles, causing light diarrhoea, which was relieved by stopping its administration [Zheng et al., 2007]. In a relatively long study (65 days), no detrimental effect on body weight was found when chitosan oligosaccharides were injected (7.1–8.6 mg/kg over 5 days). An increase in lysozyme activity was apparent on the first day post injections [Hirano et al., 1991]. However, 50 mg/kg intravenously administered chitosan caused death, probably due to blood aggregation [Hirano et al., 1988].

Self-assembled nanoparticles, formed by polymeric amphiphiles, have been demonstrated to accumulate in solid tumors due to enhanced permeability and retention effect following intravenous administration. Irrespective of the dose, a negligible quantity of self-aggregates was found in heart and lung, whereas a small amount (3.6–3.8% of dose) was detected in liver for 3 days after intravenous injection of self-aggregates. The distributed amount of self-aggregates gradually increased in tumor as blood circulation time increased. When self-aggregates loaded with doxorubicin were administered into the tumor-bearing mice via the tail vein, they exhibited lower toxicity but comparable anti-tumor activity in comparison to free doxorubicin [Parka et al., 2006].

A chitosan derivative micelle system was developed for delivering a novel anti-tumor drug, gambogic acid (GA). The physicochemical and pharmaceutical properties of GA-loaded micelles (GA-M) were evaluated and compared with the formulation GA-L-arginine (GA-L) injection. This preparation has entered phase I clinical trials. Biodistribution study indicated that ~67% of GA in the GA-M was distributed in the liver, while the value of the GA-L group was ~55%. Additionally, GA amount in the kidney was greatly reduced in the GA-M group. Also, GA-M was shown to reduce the acute toxicity after i.v. administration in mice compared with GA-L. The study indicated that GA was rapidly eliminated from the blood and transferred to the tissues, especially the liver. Moreover, GA acute toxicity and vein irritation were decreased [Qu et al., 2009a]. The study conducted by Hagenaars et al. [2010] showed that both whole inactivated influenza virus (WIV) and WIV adjuvanted with N,N,N-trimethylchitosan (TMC) formulations induced minimal local toxicity. These results provided more insight into the safety of TMC and justifed further research to develop TMC-adjuvanted nasal vaccines. Table 11 summarizes the *in vitro* and *in vivo* data of safety and toxicological studies of different chitosan derivatives.

Chapter 12

CONCLUSION

The availability of chitosan in abundance and its physiochemical properties have made it a versatile polymer for use in pharmaceutical sciences. The amino groups present in chitosan make it amenable to cross-linking with anionic polymers. These cross-linked molecules are insoluble in water. Further, trimethyl, acyl, alkyl, carboxymethyl, cyclodextrin or grafted derivatives of chitosan can be made, which possess unique property of mucoadhesivity, biocompatibility or perturbing the paracellular pathway. Chitosan being biodegradable has been found to be suitable for use in tissue engineering. To improve the healing process, chitosan has been combined with a variety of modified materials such as growth factors, extracellular matrix components and antibacterial agents. The novel properties of chitosan make it one of the most promising biopolymers for cell therapy, tissue engineering and gene therapy. It is hoped that these diverse approaches for regenerative medicine will translate from research laboratories to the clinical usage in the future. The relatively safe toxicity profile of chitosan and majority of its derivatives suggest a great promise for using it to develop a wide variety of pharmaceutical dosage forms with tailor made properties.

However, it is important to note that the issue of residual solvents should be critically addressed while using chitosan products obtained through catalytic reactions. In addition, long term toxicity, especially of products using modified chitosan molecules for parenteral use needs to be ruled out. Nevertheless, chitosan possesses a great potential for further development as a dosage form ingredient on the basis of the promising reports of dosage form performance as reviewed in this article.

Table 11. Safety indicators of chitosan derivatives

Chitosan derivatives	Particular (*in vitro / in vivo*)	Dose range	Formulation	Result	Reference
Chitosan oligomers (produced by depolymerization of chitosan)	Mouse	1000-10000 mg/kg		No deaths/ clinical toxicity signs	Qin et al., 2006
N-acetylglucosamine (chitosan's copolymer)	Rats (males and females)	0%, 0.625%, 1.25%, 2.5%, 5%	Dietary administration	No obvious toxicity observed	Lee et al., 2004a
Glucosamine derivative of chitosan	Rats	300-2700 mg/kg/day	Dietary administration	No adverse effects	Anderson et al., 2005
	Dogs	159-2149 mg/Kg	Dietary administration	No adverse effects	
Chitosan oligomers (produced by depolymerization of chitosan)	Male and female rats	0, 750, 1500, 3000 mg/Kg/Day	Dietary administration	No signs of toxicity	Qin et al., 2006
Oligo-N-glucosamine (consists of the monomer and oligomers of N-acetyl- D- glucosamine)	Male and female rats	641 or 3640 mg/kg/Day	Dietary administration	No obvious toxixity, findings were limited to occasional increase in food consumption	Tago et al., 2007
Biocompatibility study of Chitosan polymers	In vitro, murine melanoma B16F10 cells	1 μm/ml-3 mg/ml	Microspheres	Cytotoxicity (IC$_{50}$ of 0.21 mg/ml range), damage to erythrocyte membrane	Carreno-Gomez and Ducan, 1997
N-[(2-hydroxy-3-trimethylammonium)propyl]chitosan chloride, further modified by introducing functional (acrylamidomethyl) groups		Concentration of 10 ppm		Complete bacterial reduction (with *Staphylococcus aureus* and *Escherichia coli*)	Lim and Hudson, 2004
Aminated chitosan	*In vitro* (for bactericidal activity)	1 ml of 1 % chitosan Derivatives		Bacteriostatic and bactericidal	Eldin et al., 2008
Aminated chitosan	Using caco 2 cell line with the direct	0.6 x 10^5 cells each were		Viability of the live cell was decreased by	Eldin et al., 2008

	connect method (for cytotoxicity)	established on Petri-dishes			
				increase the amine group substitution, however, the toxicity of aminated chitosan is negligible.	
N-trimethyl chitosan	Intestinal Caco-2 cell monolayers	Polymers of different degrees of substitution (20, 40 and 60%) at 1.0% (w/v concentration).		No substantial cell membrane damage	Thanou et al., 1999
N,N,N-trimethylchitosan				Minimal local toxicity	Hagenaars et al., 2010
Glycol chitosans	Mice		Nanoparticles	Lower toxicity than but comparable anti-tumor activity to free doxorubicin	Parka et al., 2006
Trimethyl chitosan		IC_{50} >10 mg/ml		Low cytotoxicity	Kean et al., 2005

REFERENCES

Aime, S., Gianolio, E., Uggerib, F., Tagliapietra, S., Barge, A. & Cravotto G. (2006). New paramagnetic supramolecular adducts for MRI applications based on non-covalent interactions between Gd (III)-complexes and β- or γ-cyclodextrin units anchored to chitosan. *J. Inorg. Biochem.,* 100, 931-938.

Akgün, S., Ekici, G., Mutlu, N., Beşirli, N. & Baki Hazer, B. (2007). Synthesis and properties of chitosan-modified poly(vinyl butyrate). *J. Polym. Res.,* 14, 215-221.

Aly, A. S. (1998). Self-dissolving chitosan. I: Preparation, characterization and evaluation for drug delivery system. *Macromolecular: Mat. Eng.,* 259, 13-18.

Anal, A. K., Stevens, W. F. & Remuñán-López. C. (2006). Ionotropic cross-linked chitosan microspheres for controlled release of ampicillin. *Int. J. Pharm.* 312, 166-173.

Anderson, J. W., Nicolosi, R. J. & Borzelleca, J. F. (2005). Glucosamine effects in humans: a review of effects on glucose metabolism, side effects, safety considerations and efficacy. *Food Chem. Toxicol.,* 43, 187–201.

Andreas, B. S., Hornof, M. & Zoidl, T. (2003). Thiolated polymers – thiomers: modification of chitosan with 2-iminothiolane. *Int. J. Pharm.,* 260, 229–237.

Aoki, N., Kinoshita, K., Mikuni, K., Nakanishi, K. & Hattori. K. (2007). Adsorption of 4-nonylphenol ethoxylates onto insoluble chitosan beads bearing cyclodextrin moieties. *J. Incl. Phenom. Macrocycl. Chem.,* 57, 237–241.

Aoki, N., Nishikawa, M. & Hattori, K. (2001). Synthesis of chitosan derivative having .Beta.-Cyclodextrin moiety. *Nippon Kagakkai Koen Yokoshu,* 79, 799-806.

Aoki, N., Nishikawa, M. & Hattori, K. (2003). Synthesis of chitosan derivatives bearing cyclodextrin and adsorption of *p*-nonylphenol and bisphenol A. *Carbohydr. Polym.,* 52, 219-223.

Artursson, P. (1997). Chitosans as absorption enhancers for poorly absorbable drugs. 2. Mechanism of absorption enhancement. *Pharm. Res.,* 14, 923–929.

Atyabi, F., Majzoob, S., Dorkoosh, F., Sayyah, M. & Ponchel, G. (2007). The impact of trimethyl chitosan on in vitro mucoadhesive properties of pectinate beads along different sections of gastrointestinal tract. *Drug Develop. Ind. Pharm.,* 33, 291 – 300.

Auzely-Velty, R. & Rinaudo, M. (2002). New Supramolecular Assemblies of a Cyclodextrin-Grafted Chitosan through Specific Complexation, *Macromolecules.* 35, 7955-7962.

Auzely-Velty, R. & Rinaudo, M. (2001). Chitosan derivatives bearing pendant cyclodextrin cavities: Synthesis and inclusion performance, *Macromol.,* 34, 3574-3580.

Avadi, M. R., Sadeghi, A. M. M., Tahzibi, A., Bayati, K., Pouladzadeh, M., Zohuriaan-Mehr, M. J. & Rafiee-Tehrani, M. (2004). Diethylmethyl chitosan as an antimicrobial agent: Synthesis, characterization and antibacterial effects. *Eur. Polym. J.,* 40, 1355-1361.

Azad, A. K., Sermsintham, N., Chandrkrachang, S. & Stevens, W. F. (2004). Chitosan membrane as a wound-healing dressing: characterization and clinical application, *J. Biomed. Mater. Res. B Appl. Biomater.,* 69, 216–222.

Badawy, M. E. I., Rabea, E. I., Rogge, T. M., Stevens, C. V., Steurbaut, W., Höfte, M. & Smagghe, G. (2005). Fungicidal and insecticidal activity of O-acyl chitosan derivatives. *Poly. Bul.,* 54, 279-289.

Bal, S. M., Bram, B., Riet, E., Kruithof, A., Ding, Z., Gideon F. A. Kersten, Jiskoot, Joke, W. & Bouwstra. A. (2009). Efficient induction of immune responses through intradermal vaccination with N-trimethyl chitosan containing antigen formulations. *J. Control. Rel., In Press,* (doi:10.1016/j.jconrel.2009.11.018).

Barrett, W. C., DeGnore, J. P., König, S., Fales, H. M., Keng, Y. F., Zhang, Z. Y., Yim, M. B. & Chock, P. B. (1999). Regulation of PTP1B via glutathionylation of the active site cysteine 215. Biochem., 38, 6699-6705.

Bernkop-Schnu¨ rch, A., Paikl, C., & Valenta, C. (1997). Novel bioadhesive chitosan–EDTA conjugate protects leucine enkephalin from degradation by aminopeptidase *N. Pharm. Res.,* 14, 917–922.

Bernkop-Schnurch, A. & Hopf, T. E. (2001). Synthesis and in vitro evaluation of chitosan–thioglycolic acid conjugates. *Sci. Pharm.,* 69, 109-118.

Bernkop-Schnurch, A., Brandt, U. M. & Clausen, A. E. (1999). Synthesis and in vitro evaluation of chitosan-cysteine conjugates. *Sci. Pharm.,* 67, 196-208.

Berscht, P. C., Nies, B. & Liebrndorfer, A. (1995). In vivo evaluation of biocompatibility of different wound dressing materials, *J. Mater. Sci. Mater. Med.,* 6, 201–205.

Bhattarai, N., Ramay, H. R., Chou, S., & Zhang, M. (2006). Chitosan and lactic acid-grafted chitosan nanoparticles as carriers for prolonged drug delivery. *Int. J. Nanomed.* 1, 181–187.

Bhattarai, S. R., Remantbahadur, K. C., Aryal, S., Khil, M. S. & Kim, H.Y. (2007). *N*-Acylated chitosan stabilized iron oxide nanoparticles as a novel nano-matrix and ceramic modification *Carbohydr. Polym.,* 69 467-477.

Bhattarai, N., Ramay, H. R., Gunn, J., Matsen, F. A., Zhang, M. (2005). PEG-grafted chitosan as an injectable thermosensitive hydrogel for sustained protein release *J. Control. Rel.,* 103, 609–624.

Binello, A., Cravotto G., Nano, G. M. and Spagliardi, P. (2004). Synthesis of chitosan-cyclodextrin adducts and evaluation of their bitter-masking properties. *Flav. Frag. J.,* 19, *394 – 400.*

Boden, S. D. (2002). Overview of the biology of lumbar spine fusion and principles for selecting a bone graft substitute. *Eur. Spine J.,* 27, 26–31.

Bolgül Y., Hekimoglu S., Sahin-Erdemli I. & Kas H. S. (1998). Evaluation of oxprenolol hydrochloride permeation through isolated human skin and pharmacodynamic effect in rats. *STP Pharm. Sci.,* 8, 197-201.

Bowman, K. & Leong, K.W. (2006). Chitosan nanoparticles for oral drug and gene delivery. *Int. J. Nanomed.,* 1, 117–128.

Bravo-Osuna, I., Vauthier, C., Farabollini, A., Millotti, G. & Ponchel. G. (2008). Effect of chitosan and thiolated chitosan coating on the inhibition behaviour of PIBCA nanoparticles against intestinal metallopeptidases. *J. Nanopart. Res.,* 10, 1293–1301.

Cafaggi, S., Russo, E., Stefani, R., Leardi, R., Caviglioli, G., Parodi, B., Bignardi, G., Totero, D., Aiello, C. & Viale, M. (2007). Preparation and evaluation of nanoparticles made of chitosan or N-trimethyl chitosan and a cisplatin–alginate complex. *J. Control. Rel.,* 121, 110-123.

Campos, M., Cordi, L., Duran, N., & Mei, L. (2006). Antibacterial activity of chitosan solution for wound dressing. *Macromol. Symp.,* 245–246, 515–518.

Carreño-Gómez, B. & Duncan, R. (1997). Evaluation of the biological properties of soluble chitosan and chitosan microspheres. *Int. J. Pharm., 148, 231–240.*

Chang, J., Liu, W., Han, B. & Liu, B. (2008). The evaluation on biological properties of carboxymethyl-chitosan and carboxymethyl-chitin. *J. Ocean Univ. China. 7, 404-410.*

Chen X.G. & Park, H.J. (2003). Chemical characteristics of *O*-carboxymethyl chitosans related to the preparation conditions. *Carbohydr. Polym.,* 53, 355-359.

Chen, C. Y., Chen, C. C & Chung, Y. C. (2007). Removal of phthalate esters by α-cyclodextrin-linked chitosan bead. *Bioresour. Technol.,* 98, 2578-2583.

Chen, H. F., Pan, S. R. & Hu, Y. (2005). Study on biodegradation of carboxymethyl chitosan with different substitution *in vitro. China Pharm.,* 8, 807-809.

Chen, L. Tian, Z. & Du, Y. (2004a). Synthesis and pH sensitivity of carboxymethyl chitosan-based polyampholyte hydrogels for protein carrier matrices. *Biomat.,* 25, 3725-3732.

Chen, L. Y., Du, Y. M. & Liu, Y. (2000). Structure-antimicrobial ability relationship of carboxymethyl. *Chitosan. J. Wuhan Univ.,* 4, 191-194.

Chen, S. & Wang, Y. (2001). Study on β-cyclodextrin grafting with chitosan and slow release of its inclusion complex with radioactive iodine. J. App. Poly. Sci., 82, *2414-2421.*

Chen, S. C., Wu, Y. C., Mi, F. L., Lin, Y. H., Yu, L. C. and Sung, H. W. (2004b). A novel pH. *J. Control. Rel.,* 96, 285-300.

Clausen, A. E., Kast C. E. & Bernkop-Schnu¨ rch, A. (2002). The role of glutathione in the permeation enhancing effect of thiolated polymers. *Pharm. Res.* 19, 602–608.

Dang, J. M., Sun, D. D. N., Shin-Ya, Y., Sieber, A. N., Kostuik, J. P. & Leong, K. W. (2006). Temperature-responsive hydroxybutyl chitosan for the culture of mesenchymal stem cells and intervertebral disk cells. *Biomat.,* 27, 406-418.

Danielsen, S., Strand, S., Davies, C. & Stokke, B. T. (2005). Glycosaminoglycan destabilization of DNA–chitosan polyplexes for gene delivery depends on chitosan chain length and GAG properties. *Biochim. Biophys. Acta,* 1721, 44-54.

David, J. W. & Samuel, M. H. (2002). Heterogeneous graft copolymerization of chitosan powder with methyl acrylate using trichloroacetyl-manganese carbonyl co-initiation. *Macromol.* 35, 3413-3419.

Deng, J., Wang, L., Lianying Liu, L. & Yang, W. (2009). Developments and new applications of UV-induced surface graft polymerizations. *Progress Polym. Sci.,* 34, 156-193.

Denkbas, E. B., Seyyal, M. & Piskin, E. (2000). Implantable 5-fluorouracil loaded chitosan scaffolds prepared by wet spinning, *J. Membr. Sci.,* 172, 33–38.

Desbrieres, J., Martinez, C. & Rinaudo, M. (1996). Hydrophobic derivatives of chitosan: Characterization and rheological behaviour. *Int. J. Biol. Macromol.,* 19, 21-28.

Ding, P. Huang, K.L. Li, G.Y. Liu, Y. F. (2007a) Preparation and properties of modified chitosan as potential matrix materials for drug sustained-release beads. *Int. J. Biol. Macromol.* 41, 125-131.

Ding, P., Huang, K. L., Li, G. Y. & Zeng, W.W. (2007b). Mechanisms and kinetics of chelating reaction between novel chitosan derivatives and Zn (II). *J. Hazard. Mater.,* 146, 58-64.

Dodane, V., Khan, A. M. & Merwin J.R. (1999). Effect of chitosan on epithelial permeability and structure. *Int. J. Pharm.,* 182, 21-32.

Don, T. M. & Chen, H. R. (2005). Synthesis and characterization of AB-crosslinked graft copolymers based on maleilated chitosan and *N*-isopropylacrylamide. *Carbohydr Polym.,* 61, 334-347.

Donges, R. Reichel, D. Birgit, K. (2000). Process for the preparation and work-up of N-hydroxyalkylchitosans soluble in aqueous medium. *U. S. Patent.* 6,090,928.

Du, J. & Hsieh. Y. (2008). Nanofibrous membranes from aqueous electrospinning of carboxymethyl chitosan. *Nanotech.,* 19, 125707.1-125707.9

Dutta, P.K., Datta, J. & Tripathi. V.S. (2004). Chitin and chitosan: Chemistry, properties and applications. *J. Sci. Ind. Res.,* 63, 20-31.

Dyer A M., Hinchcliffe, M., Watts, P., Castile, J., Jabbal-Gill, I., Nankervis, R., Smith, & Illum, L. (2002). Nasal delivery of insulin using novel chitosan based formulations: a comparative study in two animal models between simple chitosan formulations and chitosan nanoparticles. *Pharm. Res.,* 19, 998-1008.

Eldin, M. S. M., Soliman, E. A., Hashem, A. I. & Tamer. T. M., (2008). Antibacterial activity of chitosan chemically modified with new technique. *Trends Biomater. Artif. Organs,* 22, 125-137.

El-Tahlawy, K. & Hudson, S. M. (2001). Graft copolymerization of hydroxyethyl methacrylate onto chitosan. *J. Appl. Poly. Sci.,* 82, 683 – 702.

El-Tahlawy, K., Gaffar, M. A. & El-Rafie. S. (2006). Novel method for preparation of β-cyclodextrin/grafted chitosan and its application. *Carbohydr. Polym.,* 63, 385-392.

Felt, O., Buri, P. & Gurny, R. (1998). Chitosan: A unique polysaccharide for drug delivery. *Drug Dev. Ind. Pharm.,* 24, 978–993.

Feng, H. & Dong, C. M. (2007). Synthesis and characterization of phthaloyl-chitosan-g-poly(l-lactide) using an organic catalyst. *Carbohydr. Polym.,* 70, 258-264.

Fu, X., Huang , L., Gao, F., Li , W., Pang , N., Zhai , M., Liu, H. & Wu, M. (2007). Carboxymethyl chitosan-coated capillary and its application in CE of proteins. *Electrophoresis,* 28, 1958 –1963.

Fujii, S., Kumagai, H. & Noda, M. (1980). Preparation of poly(acyl) chitosans. *Carbohydr. Res.,* 83, 389-393.

Gaffar, M. A., El-Rafie, S. M. & El-Tahlawy, K.F. (2004) Preparation and utilization of ionic exchange resin via graft copolymerization of β-CD itaconate with chitosan. *Carbohydr. Polym.,* 56, 387-396.

Georgeta, M., Elie, A. J., Didier, L., Luc, P., Adrian, C. & Guy M. (2004). Synthesis of chitosan microspheres containing pendant cyclodextrin moieties and their interaction with biological active molecules. *Curr. Drug Deliv.,* 1, 227-233.

Grafstrom, R., Stead, A. H. & Orrenius, S. (1980). Metabolism of extracellular glutathione in rat small-intestinal mucosa. *Eur. J. Biochem.,* 106, 571–577.

Grant, S., Blair, H. S. & Mckay, G. (1990). Deacetylation effects on the dodecanoyl substitution. *Polym. Commun.,* 31, 267-268.

Green, S., Roldo, M., Douroumis, D., Bouropoulos, N., Lamprou D. &. Fatouros. D.G. (2009). Chitosan derivatives alter release profiles of model compounds from calcium phosphate implants. *Carbohydr. Res.,* 344, 901-907.

Grobouillot, A. R., Champagne, C. P., Darling, G. D., Poncelet, D. & Neufeld, R.J. (1993). Membrane formation by interfacial cross-linking of chitosan for microencapsulation of *Lactococcus lactis. Biotechnol. Bioeng.* 42, 1157-1163.

Gruber, J. V. (1997). Oxirane carboxylic acid derivatives of polyglucosamines. *U.S. Patent.* 5,597,811.

Gruber, J. V., Venceslav, R. P., Bandekar, J. & Konish, P. N. (1995). Synthesis of N-[(3'-Hydroxy-2',3'-dicarboxy)- ethyl]chitosan: *A New, Water-Soluble Chitosan Derivative. Macromol.,* 28, 8865-8867.

Guangyuan, L., Baiyang, S., Gan, W., Yujun, W., Yandao, G., Xiufang, Z. & Lihai, Z. (2009). Controlling the degradation of covalently cross-linked carboxymethyl chitosan utilizing bimodal molecular weight distribution. *J. Biomat. Appl.*, 23, 435-451.

Guggi, D., Krauland, A. H. & Bernkop-Schnürch, A. (2003). Systemic peptide delivery via the stomach: in vivo evaluation of an oral dosage form for salmon calcitonin. *J. Control. Rel.*, 92, 125-135.

Gupta, K. C. & Kumar, R. (2000). An overview on chitin and chitosan applications with an emphasis on controlled drug release formulations. *Polymer Rev.*, 40, 273-308.

Hagenaars, N., Mania, M., Jong, P., Que, I., Nieuwland, R., Slütter, B., Glansbeek, H., Heldens, J., Bosch, H., Löwik, C., Kaijzel, E., Mastrobattista, E. & Jiskoot. W. (2010). Role of trimethylated chitosan (TMC) in nasal residence time, local distribution and toxicity of an intranasal influenza vaccine. *J. Control. Rel.*, In Press, (doi:10.1016/j.jconrel.2010.01.027)

Hamman, J. H., Schultz, C. M. & Kotze', A. F. (2003). N-Trimethyl chitosan chloride: Optimum degree of quaternization for drug absorption enhancement across epithelial cells. *Drug Dev. Ind. Pharm.*, 29, 161-172.

Harris, J. M., Struck, E. C., Case, M. G., Paley, M. P., Yalpani, M., Van Alstine, J. M., & Brooks, D. E. (1984). Synthesis and characterization of poly(ethylene glycol) derivatives, *J. Polym. Sci. Part A Polym. Chem.*, 22, 341–352.

Hassan, E. E. & Gallo, J. M. (1990). A simple rheological method for the in vitro assessment of mucin polymer bioadhesive bond strength. *Pharm. Res.*, 7, 491-495.

Hirano, S. & Midorikawa, T. (1998). Novel method for the preparation of N-acylchitosan fiber and N-acylchitosan-cellulose fiber. *Biomat.*, 19, 293-297.

Hirano, S. & Moriyasu, T. (1981). *N*-(Carboxyacyl) chitosans. *Carbohydr. Res.*, 92 323-327.

Hirano, S. & Moriyasu, T. (2004). Some novel *N*-(carboxyacyl) chitosan filaments. *Carbohydr. Polym.*, 55, 245-248.

Hirano, S. & Ohe, Y.Y. (1975) Chitosan gels: a novel molecular aggregation of chitosan in acidic solutions on a facile acylation. *Agric. Biol. Chem.*, 39, 1337-1338.

Hirano, S. Ohe, Y. & Ono, H. (1976). Selective *N*-acylation of chitosan. *Carbohydr. Res.*, 47, 315-320.

Hirano, S. Yamaguchi, Y. & Kamiya, M. (2002). Novel *N*-saturated-fatty-acyl derivatives of chitosan soluble in water and in aqueous acid and alkaline solutions. *Carbohydr. Polym.,* 48, 203-207.

Hirano, S., Iwata, M., Yamanaka, K., Tanaka, H., Toda, T. & Inui, H. (1991). Enhancement of serum lysozyme activity by injecting a mixture of chitosan oligosaccharides intravenously in rabbits. *Agric. Biol. Chem.* 55, 2623–2625.

Hirano, S., Seino, H., Akiyama, Y. & Nonaka, I. (1988). Bio-compatibility of chitosan by oral and intravenous administrations, *Proceedings of the ACS Division of Polymeric*

Hirano, S., Zhang, M., Chung, B. & Kim, S. K (2000). The *N*-acylation of chitosan fibre and the *N*-deacetylation of chitin fibre and chitin–cellulose blended fibre at a solid state. *Carbohydr. Polym.,* 41, 175-179.

Hoffman, A.S., Chen, G. Wu, X. Ding, Z. Kabra, B. Randeri, K. Schiller M., Ron, E.S. Peppas N.A.& Brazel. C.S. (1997). Graft copolymers of peo-ppo-peo triblock polyethers on bioadhesive polymer backbones: synthesis and properties. *Polym. Preprints,* 38, 524-525

Holappa, J., Nevalainen, T., Soininen, P., Másson, M. & Järvinen, T. (2006). Synthesis of novel quaternary chitosan derivatives via *N*-Chloroacyl-6-*O* triphenylmethylchitosans. *Biomacromol.,* 7, 407–410.

Hoshino, M., Namikawa, T., Kato, M., Terai, H., Taguchi, S., & Takaoka, K. (2007). Repair of bone defects in revision hip arthroplasty by implantation of a new boneinducing material comprised of recombinant human BMP-2, Beta-TCP powder, and a biodegradable polymer: An experimental study in dogs, *J. Orthop. Res.,* 25, 1042–1051.

Hu, S. G., Jou, C. H., & Yang, M.C. (2003). Antibacterial and biodegradability properties of polyhydroxyalkanoates grafted with chitosan and chitooligosaccharides via ozone treatment, *J. Appl. Polym. Sci.,* 88, 2797–2803.

Hu, Y., Ding, Y., Ding, D., Sun, M., Zhang, L., Jiang, X. & Yang, C. (2007a) Hollow chitosan/poly (acrylic acid) nanospheres as drug carriers. *Biomacromol.,* 8, 1069–1076.

Hu, Y., Du, Y., Yang, J., Tang, Y., Li, J. & Wang, X. (2007b). Self-aggregation and antibacterial activity of *N*-acylated chitosan. *Polym.,* 48, 3098-3106.

Huang, M., Shen, X., Sheng, Y. & Fang, Y. (2005). Study of graft copolymerization of *N*-maleamic acid-chitosan and butyl acrylate by γ-ray irradiation. *Int. J. Biol. Macromol.,* 36, 98-102.

Huh, K. M., Cho, Y. W., Chung, H., Kwon, I. C., Jeong, S. Y., Ooya, T. & Lee, W. K. (2004). Supramolecular hydrogel formation based on inclusion complexation between poly(ethylene glycol)-modified chitosan and a-cyclodextrin. *Macromol. Biosci.,* 4, 92–99.

Huh, M. W., Kang, I. K., Lee, D. H., Kim, W. S., Lee, D. H. & Park, L. S. (2001). Preparation and antibacterial activity of PET chitosan nanofibrous mats using an electrospinning technique. *J. Appl. Polym. Sci.,* 81, 2769–2778.

Ieva, E., Trapani, A., Cioffi, N., Ditaranto, N., Monopoli, A. Sabbatini. L. (2009). Analytical characterization of chitosan nanoparticles for peptide drug delivery applications. *Anal. Bioanal. Chem.,* 393, 207–215.

Ikada, Y. (1984). Blood-compatible polymers, In *Adv. Polym. Sci.,* Springer Link, 57, 103–115.

Illum, L. (1998). Chitosan and its use as a pharmaceutical excipients. *Pharm. Res.,* 15, 1326-1331.

Illum, L., Farraj, N. F. & Davis, S. S. (1994). Chitosan as a novel nasal delivery system for peptide drugs. *Pharm. Res.,* 11, 1186-1189.

Illum, L., Watts, P., Fisher, A. N., Gill, I. J. & Davis, S. S. (2000). Novel chitosan-based delivery systems for the nasal administration of a LHRH analogue. *STP, Pharm. Sci.,* 10, 89–94.

Jayakumar, R. Reis, R. L. & Mano, J. F. (2006) Synthesis of N-Carboxymethyl Chitosan Beads for Controlled Drug Delivery Applications. *Mater. Sci. Forum.* 514–516, 1015–1019.

Jayakumar, R., New, N., Tokura, S. & Tamura H. (2007). Sulfated chitin and chitosan as novel biomaterials. *Int. J. Biol. Macromol.,* 40, 175–181.

Je, J. Y. & Kim, S. K. (2005). Water-soluble chitosan derivatives. *Bioorg. Med. Chem.,* 13, 6551-6555.

Je, J. Y. & Kim, S. K. (2006). Reactive oxygen. *Bioorg. Med. Chem.,* 14, 5989-5994.

Jenkins D. W. & Hudson, S. M. (2001). Review of vinyl graft copolymerization featuring recent advances toward controlled radical-based reactions and illustrated with chitin/chitosan trunk polymers. *Chem. Rev.,* 101, 3245–3274.

Jeong, Y. I., Kim, S. H., Jung, T. Y., Kim, I. Y., Kang, S. S., Jin, Y. H., Ryu, H. H. Sun, H. S., Jin, S., Kim, K. K., Ahn, K.Y. & Jung, S. (2006). Polyion complex micelles composed of all-trans retinoic acid and poly (ethylene glycol)-grafted-chitosan. *J. Pharm. Sci.,* 95, 2348–2360.

Jia, Z., Shen, D. & Xu, W. (2001). Synthesis and antibacterial activities of quaternary ammonium. *Carbohydr. Res.,* 333, 1-6.

Jiang, G., Quan, D., Kairong Liao, K. & Wang, H. (2006). Novel polymer micelles prepared from chitosan grafted hydrophobic palmitoyl groups for drug delivery. *Mol. Pharm.,* 3, 152-160.

Joshi, J. M. & Sinha, V. K. (2007). Ceric ammonium. *Carbohydr. Polym.* 67, 427-435.

Jou, C. H., Lin, S. M., Hwang, M. C., Yu, D. G., Chou, W. L., Lee, J. S., & Yang, M. C. (2007). Biofunctional properties of polyester fibers grafted with chitosan and collagen, *Polym. Adv. Technol.,* 18, 235–239.

Jumaa, M., Furkert, F. H. & Muller, B.W. (2002). A new lipid emulsion formulation with high antimicrobial efficacy using chitosan, *Eur. J. Pharm. Biopharm.,* 53, 115–123.

Jung, B.O., Kim, C. H., Choi, K. S., Lee, Y. M. & Kim, J. J. (1999). Preparation of amphiphilic chitosan and their antimicrobial activities. *J. Appl. Polym. Sci.,* 72, 1713-1719.

Kafedjiiski, K., Krauland, A. H., Hoffer, M. H. and Bernkop-Schnürch, A. (2005). Synthesis and in vitro evaluation of a novel thiolated chitosan. *Biomat.,* 26, 819-826.

Kamel, A., Sokar, M., Naggar, V. & Gamal S. (2002). Chitosan and sodium alginate-based bioadhesive vaginal tablets. *AAPS PharmSci.,* 4, 224-230.

Kang, H. M., Cai, Y. L. & Liu, P. S. (2006). Synthesis, characterization and thermal sensitivity of chitosan-based graft copolymers. *Carbohydr. Res.,* 341, 2851-2857.

Kast, C. E. & Bernkop-Schnurch, A. (2001). Thiolated polymers-thiomers: development and in vitro evaluation of chitosan–thioglycolic acid conjugates. *Biomat.,* 22, 2345-2352.

Kast, C. E., Valenta, C., Leopold, M. & Bernkop-Schnurch, A. (2002). Design and in vitro. *J. Control. Rel.,* 81, 347-354.

Kast, C.E. & Andreas, B. S. (2001). Thiolated polymers – thiomers: development and in vitro evaluation of chitosan-thioglycolic acid conjugates. *Biomat.,* 22, 2345–2352.

Kato, Y. Onishi, H. & Machida, Y. (2001a). Biological characteristics of lactosaminated *N*-succinyl-chitosan as a liver. *J. Control. Rel.,* 70, 295-307.

Kato, Y. Onishi, H. & Machida, Y. (2001b). Lactosaminated and intact N-succinyl-chitosans as drug carriers in liver metastasis. *Int. J. Pharm.,* 226, 93-106.

Kato, Y., Onishi, H. & Machida, Y. (2002). Depolymerization of *N*-succinyl-chitosan by hydrochloric acid. *Carbohydr. Res.,* 337 561-564.

Kato, Y., Onishi, H. & Machida, Y. (2004). *N*-succinyl-chitosan as a drug carrier: water-insoluble and water-soluble conjugates. *Biomat.,* 25, 907-915.

Kean, T., Roth, S. & Thanou, M. (2005). Trimethylated chitosans as non-viral gene delivery vectors: cytotoxicity and transfection efficiency, *J. Control. Rel.,* 103, 643–653.

Kaur, G., Jain, S. and Tiwary, A. K. (2009a). Chitosan-carboxymethyl tamarind kernel powder interpolymer complexation: investigations for colon drug delivery. *Sci Pharm.* ,doi:10.3797/scipharm.0908-10.

Kaur, G., Rana, V., Jain, S. and Tiwary, A. K. (2009b). Colon Delivery of Budesonide: Evaluation of Chitosan–Chondroitin Sulfate Interpolymer Complex. *AAPS Pharm. Sci. Tech.* Doi: 10.1208/s12249-009-9353-8.

Khnor, E. & Lim, L. (2003) Implantated applications of chitin and chitosan. *Biomat.,* 24, 339–49.

Kim, C. H., Choi, J. W., Chun, H. W.& Choi, K. S. (1997). Synthesis of chitosan derivatives with quaternary ammonium salt and their antibacterial activity. *Poly. Bul.,* 38, 387-393.

Kim, I. Y., Kim, S. J., Shin, M. S., Lee, Y. M., Shin, D. I. & Kim, S. I. (2002). pH- and thermal characteristics of graft hydogels based on chitosan and poly(dimethylsiloxane). *J. Appl. Polym. Sci.,* 85, 2661–2666.

Kim, J. H., Kim, Y. S., Kim, S. W., Park, J. H., Kim, K., Choi, K., Chung, . H., Jeong, S. Y., Park, R. W., Kim, I. S. & Kwon, I.C. (2006). Hydrophobically modified glycol chitosan nanoparticles as carriers for paclitaxel. *J. Control. Rel.,* 111, 228-234.

Kim, S. Y., Cho, S. M., Lee, Y. M. & Kim, S. J. (2000). Thermo- and pH-responsive behaviors of graft copolymer and blend based on chitosan and N-isopropylacrylamide. *J. Appl. Polym. Sci.,* 78, 1381–1391.

Kim, T. H., Ihm, J. E., Cho, Y. J., Nah, J. W. & Cho, C. S. (2003). Efficient gene. *J. Control. Rel.,* 93, 389-402.

Kim, Y.H., Gihm, S. H., Park, C. R., Lee, K. Y., Kim, T. W., Kwon, I. C., Chung, H. & Jeong S. Y. (2001). Structural characteristics of size-controlled self-aggregates. *Bioconjugate Chem.* 12, 932-938.

Kogan, G., Skorik, Y. A., Zjitnanova, I., Krizkova, L., Djrackova, Z., Gomes C. A. R., Yatluk, Y. G. & Krajcovic, J. (2004). Antioxidant and antimutagenic activity of *N*-(2-carboxyethyl) chitosan. *Toxicol. Appl. Pharmacol.* 201, 303-310.

Kotze', A. F., Leeuw, B. J., Lueßen, H. L., Boer, A. G., Verhoef J.C. & Junginger, H. E. (1997b). Chitosans for enhanced delivery of therapeutic peptides across intestinal epithelia: in vitro evaluation of Caco-2 cell monolayers. *Int. J. Pharm.,* 159, 243–253.

Kotze´, A. F., Lueßen, H. L., Leeuw, B. J., Boer, A. G., Verhoef, J. C.& Junginger, H.E. (1998). Comparison of the effect of different chitosan salts and N-trimethyl chitosan chloride on the permeability of intestinal epithelial cells (Caco-2). *J. Control. Rel.,* 51, 35-46.

Kotze´, A. F., Lueßen, H. L., Leeuw, B. J., Boer, A. G., Verhoef, J. C.& Junginger, H.E. (1997a). N-trimethyl chitosan chloride as a potential absorption enhancer across mucosal surfaces: in vitro evaluation in intestinal epithelial cells (Caco-2). *Pharm. Res.,* 14, 1197–1202.

Kotze´, A. F., Lueßen, H. L., Thanou, M. M., Verhoef, J. C., Boer, A. G. Junginger, H. E. & Lehr, C. M. (1999a). Chitosan and chitosan derivatives as absorption enhancers for peptide drugs across mucosal epithelia, in: E. Mathiowitz, D.E. Chickering, C. M. Lehr (Eds.), Bioadhesive Drug Delivery Systems: Fundamentals, Novel Approaches and Development, Marcel Dekker, New York, pp. 341–387.

Kotze´, A. F., Thanou, M. M., Luessen, H. L., Boer, A. G.de., Verhoef, J. C.& Junginger, H. E. (1999b). Enhancement of paracellular drug transport with highly quaternized *N*-trimethyl chitosan chloride in neutral environments: in vitro evaluation in intestinal epithelial cells (Caco-2). *J. Pharm. Sci.,* 88, 253–257.

Kotzé, A. F., Thanou, M. M., Luessen, H. L., de Boer, B. G., Verhoef, J. C. & Junginger, H. E. (1999c). Effect of the degree of quaternization of N-trimethyl chitosan chloride on the permeability of intestinal epithelial cells (Caco-2). *Eur. J. Pharm. Biopharm.,* 47, 269-274.

Krauland, A. H. & Alonso, M. (2007). Chitosan/cyclodextrin nanoparticles as macromolecular drug delivery system. *Int. J. Pharm.,* 340, 134–142.

Krauland, A. H. Guggi, D. Bernkop-Schnurch, A. (2004). Oral insulin delivery: the potential of thiolated chitosan-insulin tablets on non-diabetic rats. *J. Control. Rel.,* 95, 547-555.

Kumar, M. N., Muzzarelli, R. A., Muzzarelli, C., Sashiwa, H. & Domb, A.J. (2004). Chitosan chemistry and pharmaceutical perspectives. *Chem. Rev.* 104, 6017–6084.

Kumar, T. R., Shanmugasundaram, N. & Babu, M. (2003). Biocompatible collagen scaffolds from a human amniotic membrane: Physicochemical and in vitro culture characteristic. *J. Biomater. Sci.,* 14, 689–706.

Kumbar, S. G. & Aminabhavi, T. M. (2003). Synthesis and characterization of modified chitosan microspheres: Effect of the grafting ratio on the controlled release of nifedipine through microspheres. *J. Appl. Polym. Sci.,* 89, 2940–2949.

Kuroyanagi, Y., Shiraishi, A., Shirasaki Y., Nakakita, N., Yasutomi, Y., Takano, Y. & Shioya, N. (1994). Development of a new wound dressing with antimicrobial delivery capability. *Wound Repair Regen.,* 2, 122-129.

Kweon, D. K. & Kang, D. W. (1999). Drug-release behavior of chitosan-g-poly(vinyl alcohol) copolymer matrix. J. Appl. Polym. Sci., 74, 458–464.

Kweon, D. K., Song, S. B. & Park, Y. Y. (2003). Preparation of water-soluble chitosan/heparin complex and its application as wound healing accelerator. *Biomat.,* 24, 1595–601.

Lagos, A. & Reyes. J. (2003). Grafting onto chitosan. I. Graft copolymerization of methyl methacrylate onto chitosan with Fenton's reagent (Fe^{2+}-H_2O_2) as a redox initiator. *J. Polym. Sci. Part B: Polymer Physics,* 26, 985-991.

Lang G, Maresch G, Birkel S. (1997) Hydroxyalkyl chitosans. In: Chitin Handbook, Muzzarelli, R. A. A., Peter, M. G, eds. (Grottammare: Atec Edizioni) pp. 61–66.

Lang, G. Gerhard, M. Hans-Rudi, L. Konard, E. Breuer, L. Hoch, D. (1989). Cosmetic compositions on the basis of alkyl-hydroxypropyl-substituted chitosan derivatives, new chitosan derivatives and processes for the production thereof. *U.S. Patent.* 4,845,204.

Lang, G., Konrad, E,. Wendel, H., Maresch, G. & Hans-Rudi, L. (1990). Cosmetic compostions based upon N-hydroxybutyl-chitosans, N-hydroxybutyl-chitosans as well as processes for the production thereof. *U.S. Patent.* 4,931,271.

Lang, G., Konrad, E., Wendel, H., Maresch, G., Hans-Rudi, J. & Titze J. (1988). Cosmetic compositions based upon N-hydroxypropyl-chitosans, new N-hyroxypropyl-chitosans, as well as processes for the production thereof. *U.S. Patent.* 4,780,310.

Langoth, N., Guggi, D., Pinter, Y. & Andreas, B. S. (2004). Thiolated chitosan: in vitro evaluation of its permeation enhancing properties. *J. Control. Rel.,* 94, 177-86.

Lee, J. K., Lim, H. S. & Kim, J. H. (2002). Cytotoxic activity of aminoderivatized cationic chitosan derivatives. *Bioorg. Med. Chem. Lett.,* 12, 2949-2951.

Lee, D., Lockey, R. & Mohapatra, S. (2006b). Folate Receptor-Mediated Cancer Cell Specific Gene Delivery Using Folic Acid-Conjugated Oligo chitosans. *J. Nanosci. Nanotechnol.,* 6, 2860-2866.

Lee, D.W., Shirley, S. A., Lockey, R. F. & Mohapatra, S. S. (2006a). Thiolated chitosan nanoparticles enhance anti-inflammatory effects of intranasally delivered theophylline. *Resp. Res.,* 7, 112.

Lee, E. S., Park, K. H., Park, I. S. & Na, K. (2007). Glycol chitosan as a stabilizer for protein encapsulated into poly(lactide-co-glycolide) microparticle. *Int. J. Pharm.,* 338, 310-315.

Lee, K. Y., Shibutani, M., Takagi, H., Arimura, T., Takigami, S., Uneyama, C., Kato, N. & Hirose, M. (2004a). Subchronic toxicity study of dietary N-acetylglucosamine in F344 rats. *Food Chem. Toxico.,* 42, 687-695.

Lee, W. Powers, K. Baney, R. (2004b). Physicochemical properties and blood compatibility of acylated chitosan nanoparticles. *Carbohydr. Polym.,* 58, 371-377.

Leitner, V. M., Walker, G. F. & Bernkop-Schnurch, A. (2003). Thiolated polymers. *Eur. J. Pharm. Biopharm.,* 56, 207-214.

Leng, W., Qin, L. & Tang, X. (2006). Chitosan and randomly methylated β-cyclodextrin combined to enhance the absorption and elevate the bioavailability of estradiol intranasally: in situ and in vivo Studies. *J. Am. Sci.,* 2, 61-65.

Li, M. & Xin, M. (2006). N,N-Dilauryl chitosan self-assembled vesicles for drug delivery. *Des. Monomers Polym.,* 9, 89-97.

Li, M., Su, S., Xin, M. & Liao, Y. (2007). Relationship between *N, N*-dialkyl chitosan monolayer. *J. Colloid Interf. Sci.,* 311, 285-288.

Li, Y., Liu, L., & Fang, F. (2003). Plasma-induced grafting of hydroxyethyl methacrylate (HEMA) onto chitosan membranes by a swelling method. *Polym. Int.,* 52, 285–290.

Li, Y., Liu, L., Shen, X. & Fang, L. (2005). Preparation of chitosan/poly(butyl acrylate) hybrid materials by radiation-induced graft copolymerization based on phthaloylchitosan. *Rad. Phy. Chem.,* 74, .297-301.

Liang, P., Zhao, Y., Shen, Q., Wang D. & Xu, D. (2004). The effect of carboxymethyl chitosan on the precipitation of calcium carbonate. *J. Cryst. Growth.,* 261, 571-576.

Lim, S. & Hudson, S. M. (2004). Synthesis and antimicrobial activity of a water-soluble chitosan derivative with a fiber-reactive group. *Carbohydr. Res.,* 339, 313-319.

Lin, Y. H., Liang, H. F. G., Chung, C. H., Chen, M. C. & Sung, H.W. (2005). Physically crosslinked alginate/N,O-carboxymethyl chitosan hydrogels with calcium for oral delivery of protein drugs. *Biomat.,* 26, 2105-2113.

Lin, Y., Chen, Q. & Luo, H. (2007). Preparation. *Carbohydr. Res.,* 342, 87-95.

Liu, C., Fan, W., Chen, X., Liu, C., Meng, X., Park & H. J. (2007a). Self-assembled nanoparticles. *Curr. Appl. Phys.,* 7S1, e125-e129.

Liu, L., Li, Y., Fang, Y., & Chen, L. (2005). Microwave-assisted graft copolymerization of ε-caprolactone onto chitosan via the phthaloyl protection method. *Carbohydr. Polym.,* 60, 351-356.

Liu, T. S., Chen, S. Y., Lin, Y. L. & Liu, D. M. (2006). Synthesis and characterization of amphiphatic carboxymethyl-hexanoyl chitosan hydrogel: Water-retention. *Langmuir,* 22, 9740-9745.

Liu, W.G., Zhang, X., Sun, S. J., Sun, G. J., Yao, K. D., Liang, D. C., Guo, G. & Zhang, J. Y. (2003). N-Alkylated chitosan as a potential nonviral vector. *Bioconjugate Chem.* 14, 782-789.

Liu, X. F., Guan, Y. L., Yang, D. Z., Li, Z. &. Yao, K. D. (2001). Antibacterial action of chitosan and carboxymethylated chitosan. *J. Appl. Polym. Sci.,* 79, 1324-1335.

Liu, Y., Cai, S., Shu, X. Z., Shelby, J. & Prestwich, G. D. (2007b). Release of basic fibroblast growth factor from a crosslinked glycosaminoglycan hydrogel promotes wound healing. *Wound Repair Regen.,* 15, 245–251.

Lorenzo, C. A., Concheiro, A., Dubovik, A. S., Grinberg, N. V., Burova, T. V. & Grinberg, V. Y. (2005). Temperature-sensitive chitosan-poly (*N*-isopropylacrylamide) interpenetrated networks with enhanced loading capacity and controlled release properties. *J. Control. Rel.,* 102, 629–641.

Lu, G. Kong, L., Sheng, B., Wang, G., Gong, Y. & Zhang, X. (2007). Degradation of covalently cross-linked carboxymethyl chitosan and its potential application for peripheral nerve. *Eur. Polym. J.,* 43, 3807–3818.

Lv, P., Bin, Y., Li, Y., Chen, R., Wang, X. & Zhao B. Studies on graft copolymerization of chitosan with acrylonitrile. (2009). *Polym.,* 50, 5675-5680.

Maculotti, K., Genta, I., Perugini, P., Imam, M., Bernkop-Schnurch, A. & Pavanetto, F. (2005). Preparation and in vitro evaluation of thiolated chitosan microparticles. *J. Microencap.,* 225, 459-470.

Maestrelli, F., Garcia-Fuentes, M., Mura, P. & Alonso, M. J. (2006). A new drug nanocarrier consisting of chitosan and hydroxypropylcyclodextrin. *Eur. J. Pharm. Biopharm.,* 63, 79-86.

Mansouri, S. Cuie, Y. Winnik, F. Shi, Q. Lavigne, P. Benderdour, M. Beaumont, E. Fernandes, J. C. (2006). Characterization of folate-chitosan-DNA nanoparticles for gene therapy, *Biomat.,* 27, 2060-2065.

Mao, C., Yuan, J., Mei, H., Zhu, A., Shen, J. & Lin, S. (2004a). Introduction of photocrosslinkable chitosan to polyethylene film by radiation grafting and its blood compatibility. *Mater. Sci. Eng.,* C 24, 479–485.

Mao, C., Zhao, W. B., Zhu, A. P., Shen, J. & Lin, S. C. (2004b). A photochemical method for the surface modification of poly(vinyl chloride) with O-butyrylchitosan to improve blood compatibility. *Process Biochem.,* 39, 1151–1157.

Mao, S. Shua, X. Unger, F. Wittmar, M. Xie, X. & Kissel, T. (2005). Synthesis, characterization and cytotoxicity of poly(ethylene glycol)-graft-trimethyl chitosan block copolymers. *Biomat.,* 26, 6343-6356.

Mao, S., Sun. & Kissel, T. (2010). Chitosan-based formulations for delivery. *Adv. Drug Deliv. Rev.,* 62, 12-27.

Maresch, G., Clausen, T. & Lang, G. (1989). Hydroxypropylation of chitosan. In: *Chitin and Chitosan*, Skjak-Bræk, G., Anthonsen, T., Sandford, P., eds. (Essex: Elsevier) pp. 389– 395.

Marie, E., Landfester, K. & Antonietti, M. (2002). Synthesis of chitosan-stabilized polymer dispersions, capsules, and chitosan grafting products via miniemulsion. *Biomacromol.,* 3, 475–481.

Martel, B. Devassine, M. Crini, G. Weltrowski, M. Bourdonneau, M. & Morcellet, M. (2001). Preparation sorption properties of a β-cyclodextrin-linked chitosan derivative. *J. Polym. Sci. Part A: Polym. Chem.,* 39 169-176.

Martin, L., Wilson, C. G., Koosha F., Tetley, L., Gray, A. I., Senel, S. & Uchegbu, I. F. (2002). The release of model macromolecules may be controlled by the hydrophobicity of palmitoyl glycol chitosan hydrogels. *J. Control. Rel.* 80, 87-100.

Martien, R., Loretz, B., Sandbichler, M. A. & Bernkop Schn"urch, A. (2008). Thiolated chitosan nanoparticles: transfection study in the Caco-2 differentiated cell culture. *Nanotechnol.,* 19, 045101

Merwe, S. M. V., Verhoef, J. C., Kotzé, A. F. & Junginger, H. E. (2004). *N*-Trimethyl chitosan chloride as absorption enhancer in oral peptide drug delivery. Development and characterization of minitablet and granule formulations. *Eur. J. Pharm. Biopharm.,* 57, 85-91.

Mi, F. L., Sung, H. W. & Shyu, S. S. (2002a). Drug release from chitosan–alginate complex beads reinforced by a naturally occurring cross-linking agent, *Carbohydr. Polym.,* 48, 61–72.

Mi, F. L., Tan, Y. C., Liang, H. F. & Sung, H. W. (2002b). In vivo biocompatibility and degradability of a novel injectable-chitosan-based implant. *Biomat.,* 23, 181–191.

Miwa, A., Ishibe, A., Nakano, M., Yamahira, T., Itai, S., Jinno, S. & Kawahara, H. (1998). Development of novel chitosan derivatives as micellar carriers of taxol. *Pharm. Res.,* 15, 1844-50.

Mizuno, K., Yamamura, K., Yano, K., Osada, T., Saeki, S., Takimoto, N., Sakurai T.Nimura, Y. (2003). Effect of chitosan film containing basic fibroblast growth factor on wound healing in genetically diabetic mice, *J. Biomed. Mater. Res.,* 64, 177–181.

Mocanu, G., About-Jaudet, E, LeCerf, D., Picton, L., Carpov, A & Muller, G. (2004). Synthesis of chitosan microspheres containing pendant cyclodextrin moieties and their interaction with biological active molecules. *Curr. Drug Del.,* 1, 227-233.

Moses, L. R., Dileep, K. J. & Sharma, C. P. (2000). Beta cyclodextrin–insulinencapsulated chitosan/alginate matrix: Oral delivery system. *J. Appl. Polym. Sci,* 75, 1089–1096.

Muzzarelli, R. A. A. (1988). Carboxymethylated chitins and chitosans. *Carbohydr. Polym.,* 8, 1-21.

Muzzarelli, R. A. A. (1989). Amphoteric derivatives of chitosan and their biological signifiance, In: *Chitin and Chitosan:* Sources, Chemistry, Biochemistry, Physical Properties, and Applications, Skja°k-Braek, G., Anthonsen, T. and Sandford, P. (eds). Elsiever Applied Science, London, pp.87–99.

Muzzarelli, R. A. A. (1997). Human enzymatic activities related to the therapeutic administration of chitin derivative. *Cell. Mol. Life Sci.,* 53, 131–140.

Muzzarelli, R. A. A., Ramos, V., Stanic, V., Dubini, B., Mattioli-Belmonte, M., Tosi, G. & Giardino, R. (1998). Osteogenesis promoted by calcium phosphate *N,N*-dicarboxymethyl chitosan. *Carbohydr. Polym.,* 36, 267-276.

Muzzarelli, R. A. A., Tanfani, F. Emanuelli, M. Mariotti S. N. (1982). *N*-(carboxymethylidene)chitosans and *N*-(carboxymethyl) chitosans: Novel chelating polyampholytes obtained from chitosan glyoxylate. *Carbohyd. Res.,* 107, 199-214.

Muzzarelli, R.A.A., Delben, F., Ilari, P. & Tomasetti, M. (1994). N-Carboxymethyl chitosan, a versatile chitin derivative. *Agro-Food Ind. High Tech.,* 5, 35-39.

Mwale, F., Iordanova, M., Demers, C. N., Steffen, T., Roughley, P., & Antonoiu, J. (2005). Biological evaluation of chitosan salts cross-linked to genipin as a cell scaffold for disk tissue engineering, *Tissue Eng.,* 11, 130–140.

Ng, L. T., Guthrie, J. T., Juan, Y. J. & Zhao, H. (2001). UV-cured natural polymer-based membrane for biosensor application. *J. Appl. Polym. Sci.,* 79, 466–472.

Nishimura, S., Miura, Y., Ren, L., Sato, M., Yamagishi, A. & Nishi, N. (1993) An Efficient Method for the Syntheses of Novel Amphiphilic Polysaccharides by Regio- and Thermoselective Modifications of Chitosan. *Chem. Lett.* 9, 1623-1626.

Nunthanid, J., Laungtana-anan, M., Sriamornsak, P., Limmatvapirat, S., Puttipipatkhachorn, S., Lim, L.Y. & Khor. E. (2004), Characterization of chitosan acetate as a binder for sustained release tablets. *J. Control. Rel.,* 99, 15-26.

Ohya, Y., Shiratani, M., Kobayashi, H., & Ouchi, T. (1994). Release behavior of 5-fluorouracil from chitosan-gel nanospheres immobilizing 5-fluorouracil coated with polysaccharides and their cell-specific cytotoxicity. *J. Macromol. Sci. Pure Appl. Chem.*, A31, 629–642.

Ono, K. Saito, Y., Yura, H. Ishikawa, K., Kurita, A., Akaike, T. & Ishihara, M. (2000). Photocrosslinkable chitosan as a biological adhesive. *J. Biomed. Mater. Res.,* 49, 289–295.

Oosegi, T., Onishi, H. & Machida, Y. (2008). Novel preparation of enteric-coated chitosan–prednisolone conjugate microspheres and in vitro evaluation of their potential as a colonic delivery system. *Eur. J. Pharm. and Biopharm.,* 68, 260-266.

Opanasopit, P., Ngawhirunpat, T., Chaidedgumjorn, A. Rojanarata, T., Apirakaramwong, A., Phongying, S., Choochottiros, C. & Chirachanchai, S. (2006). Incorporation of camptothecin into N-phthaloyl chitosan-g-mPEG self-assembly micellar system, *Eur. J. Pharm. Biopharm.* 64, 269–276.

Parka, J. H., Kwona, S., Lee, M., Chunga, H., Kimc, J. H., Kimc, Y. S., Parkc, R. W., Kimc, I. S., Seod, S. B., Kwona, I. C. & Jeong, S. Y. (2006). Self-assembled nanoparticles based on glycol chitosan bearing hydrophobic moieties as carriers for doxorubicin: In vivo biodistribution and anti-tumor activity. *Biomat.,* 27, 119–126.

Patel, J. K. Patel, R. P., Amin, A. F. Patel, M. M. & Patel, S. K. (2005). Formulation and Evaluation of Mucoadhesive Glipizide Microspheres. *AAPS PharmSci,* 6, E49-E55.

Pavlov, G. M. Korneeva, E. V. Harding S. E. & Vichoreva, G.A. (1998). Dilute solution properties of carboxymethylchitins in high ionic-strength solvent. *Polymer,* 39, 6951-6961.

Peng, X. & Zhang, L. (2005). Surface fabrication. *Langmuir,* 21, 1091-1095.

Peng, Y., Han, B., Liu, W. & Xu, X. (2005). Preparation and antimicrobial activity of hydroxypropyl chitosan. Carbohydr. Res. 340, 1846-1851.

Pengfei, L., Maolin, Z. & Jilan W. (2001). Study on radiation. *Rad. Phy. Chem.,* 61, 149-153.

Piyakulawat, P., Praphairaksit, N., Chantarasiri, N. & Muangsin, N. (2007). Preparation and evaluation of chitosan/carrageenan beads for controlled release of sodium diclofenac. *AAPS PharmSciTech.,* 8(4), Article 97.

Pourjavadi, A., Mahdavinia, G. R., & Zohuriaan-Mehr, M. J. (2003). Modified chitosan. II. H-chitoPAN, a novel pH-responsive superabsrobent hydrogel. *J. Appl. Polym. Sci.,* 90, 3115–3121.

Prabaharan, M. & Gong, S. (2008). Novel thiolated carboxymethyl chitosan-g-β-cyclodextrin as mucoadhesive hydrophobic drug delivery carriers. *Carbohyd. Poly.,* 73, 117-125.

Prabaharan, M. & Mano, J. F. (2006) Chitosan derivatives bearing cyclodextrin cavities as novel adsorbent matrices. *Carbohydr. Polym.* 63, 153-166.

Prabaharan, M. and Mano, J. F. (2005). Chitosan-based particles as drug delivery systems. *Drug Deliv.,* 12, 41-57.

Prabaharan, M., Rodriguez-Perez, M. A., Saja, J. A. & Mano, J. A. (2006). Preparation and characterization of poly (L-lactic acid)-chitosan hybrid scaffolds with drug release capability, *J. Biomed. Mater. Res. Part B Appl. Biomater.,* 81, 427–434.

Pujals, G., Sune-Negre, J. M. P., Garcia, P. E., Portus, M., Tico, J. R., Minarro, M. & J. Carrio. (2008). In vitro evaluation of the effectiveness and cytotoxicity of meglumine antimoniate microspheres produced by spray drying against Leishmania infantum, *Parasitol. Res.* 102, 1243–1247.

Puttipipatkhachorn, S., Nunthanid, J., Yamamoto, K. & Peck. G. E. (2001). Drug physical state and drug–polymer. *J. Control. Rel.,* 75, 143-153.

Qin, C., Gao, J., Wang, L., Zeng, L. & Liu, Y. (2006). Safety evaluation of short-term exposure to chitooligomers from enzymic preparation. *Food Chem. Toxicol.,* 44, 855–861.

Qu, G., Yao, Z., Zhang, C., Wu, X. & Ping, Q. (2009b). PEG conjugated N-octyl-O-sulfate chitosan micelles for delivery. *Eur. J. Pharm. Sci.,* 37, 98–105.

Qu, G., Zhu, X., Zhang C. & Ping, Q. (2009a). Modified chitosan derivative micelle system for natural anti-tumor product gambogic acid delivery. *Drug Del.* 16, 363-370.

Remantbahadur, K. C., Aryal, S., Bhattarai, S., Bhattarai, N., Kim, C. H. & Kim, H.Y. (2006). Stabilization of gold nanoparticles by hydrophobically-modified polycations. *J. Biomater. Sci. Polym. Ed.,* 17, 579-589.

Riske, F., Schroeder, S., Belliveau, J., Kang, X., Kutzko, J. & Menon, M. K. (2007). The use of chitosan as a flocculant in mammalian cell culture dramatically improves clarification throughput without adversely impacting monoclonal antibody recovery. *J. Biotech.,* 128, 813-823.

Roberts, G.A.F. K.E. Taylor, Synthesis, Characterization and Theory of Polymeric Network and Gels. In: C.J. Brine, P.A. Sandford, J.P. Zikakis (Eds.), *Chitin and Chitosan,* Elsevier, London and New York, 1992, 5th edition, pp. 179–183.

Rodrigues, M. R. (2005). Synthesis and investigation of chitosan derivatives formed by reaction with acyl chlorides. *J. Carbohyd. Chem.,* 24, 41–54.

Romaškevič, T., Budrienė, S., Liubertienė, A., Gerasimčik, I., Zubrienė, A. & Dienys. G. (2007). Synthesis of chitosan-*graft*-poly(ethylene glycol) methyl ether methacrylate copolymer and its application for immobilization of maltogenase. *Chemija,* 18, 33–38.

Ronghua, H., Yumin, D. & Jianhong, Y. (2003). Preparation and anticoagulant activity of carboxybutyrylated hydroxyethyl chitosan sulfates. *Carbohydr. Polym.,* 51, 431-438.

Rout, D. K., Pulapura, S. K. & Gross, R.A. (1993). Gel-sol transition and thermotropic behavior of a chitosan derivative in lyotropic solution. *Macromol.,* 26, 6007-6010.

Sadeghi, A. M. M., Dorkoosh, F. A., Avadi, M. R., Weinhold, M. B. A., Delie, F. Gurny, R., Larijani, B., Rafiee-Tehrani, M. & Junginger H. E. (2008). Permeation enhancer effect of chitosan and chitosan derivatives: Comparison of formulations as soluble polymers and nanoparticulate systems on insulin absorption in Caco-2 cells. *Eur. J. Pharm. Biopharm.,* 70, 270-278.

Saito, H., Wu, X., Harris, J. M. & Hoffman, A. S. (1997). Graft copolymers of poly (ethylene glycol) (PEG) and chitosan. *Macromol. Rapid Commun.,* 18, 547–550.

Saito, N., Okada, T., Toba, S., Miyamoto, S. & Takaoka, K. (1999). New synthetic absorbable polymers as BMP carriers: Plastic properties of poly-D,L-lactic acid-polyethylene glycol block copolymers. *J. Biomed. Mater. Res.,* 47, 104–110.

Sajomsang, W., Gonil, P. & Saesoo, S. (2009). Synthesis and antibacterial activity of methylated *N*-(4-*N,N*- dimethylaminocinnamyl) chitosan chloride. *Eur. Poly. J.,* 45, 2319-2328.

Sashiwa, H. & Shigemasa Y. (1999). Chemical modification of chitin and chitosan 2: preparation and water soluble property of *N*-acylated or *N*-alkylated partially deacetylated chitins. *Carbohydr. Polym.,* 39, 127–138.

Sashiwa, H., Kawasaki, N., Nakayama, A., Muraki, E., Yamamoto, N. & Aiba, S. (2002b). Chemical modification of chitosan. 14:[1] Synthesis of water-soluble chitosan derivatives. *Biomacromol.*, 3, 1126-1128.

Sashiwa, H., Kawasaki, N., Nakayama, A., Muraki, E., Yamamoto, N., Zhu, H., Nagano, H., Omura, Y., Saimot, H., Shigemasa, Y. & Aiba, S. (2002a). Chemical modification of chitosan. 13.[1] synthesis. *Biomacromol.*, 3, 1120-1125.

Schipper, N. G. M., Olsson, S., Hoogstraate, J. A., Boer, A. G., Varum, K. M. & Artursson, P. (1997) Chitosans as absorption enhancers for poorly absorbable drugs. 2. Mechanism of absorption enhancement, *Pharm. Res.*, 14, 923–929.

Schipper, N. G., Varum, K. M. & Artursson, P. (1996). Chitosans as absorption enhancers for poorly absorbable drugs. 1: influence of molecular weight and degree of acetylation on drug transport across human intestinal epithelial (caco-2) cells. *Pharm. Res.* 13, 1686–1692.

Schipper, N. G., Varum, K. M., Stenberg, P., Ocklind, G., Lennernas, H. & Artursson, P. (1999). Chitosans as absorption enhancers of poorly absorbable drugs. 3: influence of mucus on absorption enhancement. *Eur. J. Pharm. Sci.*, 8, 335–343.

Seo, H. & Kinemura, Y. (1989b). In: *Chitin and Chitosan*, Skjak-braek, G., Anthonsen, T., Sandford, P. (Eds.), Elsevier Applied Sciences, London, pp. 585–588.

Seo, T., Hagura, S., Kanbara, T. & Iijima, T. (1989a). Interaction of dyes with chitosan derivatives. *J. Appl. Polym. Sci.*, 37, 3011-3027.

Seo, T., Ikeda, Y., Torada, K., Nakata, Y. & Shimomura, Y. (2001). Synthesis of N,O-acylated chitosan and its sorptivity. *Chitin Chitosan Res.*, 7, 212-216

Seo, Y. Ohtake, H. Unishi, T. Iijima, T. (1995). Permeation of solutes through chemically modified chitosan membranes. *J. Appl. Polym. Sci.*, 58, 633-644.

Shigemasa, Y., Usui, H., Morimoto, M., Saimoto, H., Okamoto, Y., Minami, S. & Sashiwa H. (1999). Chemical modification of chitin and chitosan 1: preparation of partially deacetylated chitin derivatives via a ring-opening reaction with cyclic acid anhydrides in lithium chloride/*N,N*-dimethylacetamide. *Carbohydr. Polym.*, 39, 237-243.

Shiu, J. C., Ho, M., Yu, S., Chao, A., Su, Y., Chen, W., Chiang, C., Wen Pin Yang, W. P. (2010). Preparation and characterization of caffeic acid grafted chitosan/CPTMS hybrid. scaffolds. *Carbohydr. Polym.*, 79, 724-730.

Sieval, A. B., Thanou, M., Kotze, A. F., Verhoef, J. C., Brussee, J. & Junginger, H. E. (1998). Preparation and NMR characterization of highly substituted*N*-trimethyl chitosan chloride. *Carbohydr. Polym.* 36, 157-165.

Singh, D. K. & Ray, A. R. (1994). Graft copolymerization of 2-hydroxyethyl methacrylate onto chitosan films and their blood compatibility. *J. Appl. Polym. Sci.,* 53 1115-1121.

Singh, D. K. & Ray, A.R. (1998). Characterization of grafted chitosan films. *Carbohydr. Polym.,* 36, 251–255.

Singh, V., Tripathi, D. N., Tiwari, A, & Sanghi, R. (2004). Microwave promoted synthesis of chitosan-graft-poly(acrylonitrile). *J. Appl. Polym. Sci.,* 95, 820 – 825.

Sinha, V. R. & Kumria, R. (2001). Polysaccharides in colon-specific drug delivery. *Int. J. Pharm.,* 224, 19-38.

Skorik, Y. A. Gomes, C. A. R., Vasconcelos, M. T. S. D. & Yatluk, Y. G. (2003). *N*-(2-Carboxyethyl) chitosans: regioselective synthesis, characterisation and protolytic equilibria. *Carbohydr. Res.,* 338, 271-276.

Snyman, D., Hamman, J. H. & Kotze′, A.F. (2003). Evaluation of the mucoadhesive properties of N-trimethyl chitosan chloride. *Drug Dev. Ind. Pharm.,* 29, 59–67.

Snyman, D., Hamman, J. H., Kotze′, J. S., Rollings, J. E. & Kotze′, A.F. (2002). The relationship between the absolute molecular weight and the degree of quaternisation of N-trimethyl chitosan chloride. *Carbohydr. Polym.,* 50, 145–150.

Song, Y., Onishi, H. & Machida, Y. (1992). Synthesis and Drug-Release Characteristics of the Conjugates of Mitomycin C with N-Succinyl-chitosan and Carboxymethyl-chitin. *Chem. Pharm. Bull.,* 40, 2822-2825.

Song, Y., Onishi, H. & Nagai, T. (1993). Conjugate ofmitomycin c with n-succinyl-chitosan: in vitro drug release properties, toxicity and antitumor activity. *Int. J. Pharm.* 98, 121–130.

Sreenivasan, K. (1998). Synthesis and preliminary studies on a β-cyclodextrin-coupled chitosan as a novel adsorbent matrix. *J. Appl. Polym. Sci.,* 69, 1051-1055.

Staddon, J. M. Herrenknecht, K. Smales, C. Rubin, L. L. (1995). Evidence that tyrosine phosphorylation may increase tight junction permeability, *J. Cell Sci.,* 108, 609–619.

Sugunan, A., Thanachayanont, C., Dutta′ J. & Hilborn. J. G. (2005). Heavy-metal ion sensors using chitosan-capped gold nanoparticles. *Sci. Tech. Adv. Mat.,* 6, 335-340.

Sun, T., Xu, P., Liu, Q., Xue, J. & Xie, W. (2003). Graft copolymerization of methacrylic acid onto carboxymethyl chitosan. *Eur. Polym. J.,* 39, 189–192.

Tago, K., Naito, Y., Nagata, T., Morimura, T., Furuya, M., Seki, T., Kato, H. & Ohara, N. (2007). A ninety-day feeding, subchronic toxicity study of oligo-Nacetylglucosamine in Fischer 344 rats. *Food Chem. Toxicol.,* 45, 1186-1193.

Tajima, M., Izume, M., Fukuhara, T., Kimura, T. & Kuroyanagi, (2000). Development of new wound dressing composed of N-succinyl chitosan and gelatin. *Y. Seitai Zairyo.* 18, 220-226.

Tan, Y. & Liu, C. (2009). Self-aggregated nanoparticles from linoleic acid modified carboxymethyl chitosan: Synthesis, characterization and application *in vitro. Colloids Surf. B., Biointerfaces.* 69, 178-182.

Tanida, F., Tojima, T., Han, S. M., Nishi, N., Tokura, S., Sakairi, N., Seino, H. & Hamada, K. (1998) Novel synthesis of a water-soluble cyclodextrin-polymer having a chitosan skeleton. *Polym.,* 39, 5261-5263.

Tanodekaew, S., Prasitsilp, M., Swasdison, S., Thavornyutikarn, B., Posthsree, T., & Pateepasen, R. (2004). Preparation of acrylic grafted chitin for wound dressing application. *Biomat.,* 25, 1453–1460.

Tasker, R. A. Connell, B. J. Ross, S. J. Elson, C. M. (1998). Development of an injectable sustained-release formulation of morphine: antinociceptive properties in rats. *Lab. Anim.,* 32, 270–275.

Thacharodi, D. & Rao, K. P. (1995). Development and in vitro evaluation of chitosan-based transdermal drug delivery systems for the controlled delivery of propanolol hydrochloride. *Biomat.,* 16, 145–148.

Thanou, M. M., Verhoef, J. C., Romeijn, S. G., Nagelkerke, J. F., Merkus, F. M. Junginger, H. E. (1999). Effects of N-trimethyl chitosan chloride, a novel absorption enhancer, on Caco-2 intestinal epithelia and the ciliary beat frequency of chicken embryo trachea. *Int. J. Pharm.,* 185, 73-82.

Thanou, M. M., Kotze, A. F., Scharringhausen, T., Luessen, H. L., de Bocr, A. G., Verhoef, J. C. & Junginger, H. E. (2000b). Effect of degree of quaternization of N-trimethyl chitosan chloride for enhanced transport. *J. Control. Rel.,* 64, 15-25.

Thanou, M. Verhoef, J. C. & Junginger, H. E. (2001c). Oral drug absorption enhancement by chitosan and its derivatives. *Adv. Drug Del. Rev.* 52, 117–126.

Thanou, M., Florea, B. I., Geldof, M., Junginger, H. E. & Borchard, G. (2002). Quaternized chitosan oligomers. *Biomat.,* 23, 153-159.

Thanou, M., Florea, B. I., Langemeÿer, M. W. E., Verhoef, J. C., & Junginger, H. E. (2000c). N-Trimethylated chitosan chloride (tmc) improves the intestinal permeation of the peptide drug buserelin in vitro (caco-2 cells) and in vivo (rats). *Pharm*17, 27-31.

Thanou, M., Nihot, M. T., Jansen, M., Verhoef, J. C. & Junginger, H.E. (2001b). Mono-N-carboxymethyl chitosan (MCC), a polyampholytic chitosan derivative, enhances the intestinal absorption of low molecular weight heparin across intestinal epithelia in vitro and in vivo. *J. Pharm. Sci.*, 90, 38–46.

Thanou, M., Verhoef, J. C. & Junginger, H. E. (2001a). Chitosan and its derivatives as intestinal absorption enhancers, *Adv. Drug Deliv. Rev.*, 50, S91–101.

Thanou, M., Verhoef, J. C., Marbach, P. & Junginger H. E. (2000a). Intestinal absorption of octreoide: N-trimethyl chitosan chloride (TMC) ameliorates the permeability and absorption properties of the somatostatin analogue in vitro and in vivo. *J. Pharm. Sci.*, 89, 951-957.

Tien, C., Lacroix, M., Ispas-Szabo, P. & Mateescu, M. A. (2003). *N*-acylated chitosan: hydrophobic matrices for controlled drug release *J. Control. Rel.*, 93, 1-13.

Toffey, A. & Glasser, W.G. (2001). Chitin Derivatives III Formation of Amidized Homologs of Chitosan. *Cellulose,* 8, 35-47.

Tojimaa, T., Katsura, H., Nishiki, M., Nishi, N., Tokura, S. & Sakairi, N. (1999). Chitosan beads with pendant α-cyclodextrin: preparation and inclusion property to nitrophenolates. *Carbohydr. Polym.*, 40, 17-22.

Tokura, S., Nishi, N., Tsutsumi, A. & Somorin O. (1983). Studies on chitin. VIII. Some properties of water soluble chitin derivatives. *Polym. J.*, 15, 485–489.

Tokura, S., Ueno, K., Miyazaki, S. & Nishi, N. (1997). Molecular weight dependent antimicrobial activity by chitosan. *Macromol. Symp.*, 120, 1–9.

Trapani, A., Sitterberg, J., Bakowsky, U. & Thomas K. (2009). The potential of glycol chitosan nanoparticles as carrier for low water soluble drugs. *Int. J. Pharm.*, 375, 97-106.

Valenta, C., Christen, B., & Bernkop-Schnu¨ rch, A. (1998). Chitosan–EDTA conjugate: A novel polymer for topical gels. *J. Pharm. Pharmacol.*, 50, 445–452.

Vasnev, V. A., Tarasov, A. I., Markova, G. D., Vinogradova, S. V. & Garkusha, O.G (2006). Degradation of chitosan and starch by 360-kHz ultrasound. *Carbohydr. Polym.*, 64, 175-184.

Venter, J. P., Kotze, A. F., Auzely-Velty, R. & Rinaudo, M. (2006) Synthesis and evaluation of the mucoadhesivity of a CD-chitosan derivative. *Int. J. Pharm.,* 313, 36-42.

Wang, L.C. Chen, X.G. Zhong D.Y. & Xu, Q.C. (2007). Study on poly(vinyl alcohol)/carboxymethyl-chitosan blend film as local drug delivery system *J. Mater. Sci. Mater. Med.,* 18, 1125-1133.

Wang, W., McConaghy, A. M., Tetley, L. & Uchegbu, I. F. (2001). Controls on polymer. *Langmuir,* 17, 631-636.

Wang, Y. Y., Zhang, Y., Pang, X., Peng, Z.& Hui, Q. L. (2005). Synthesis of Novel Chitosan Microspheres Grafted with β-Cyclodextrins and Their Adsorption for Iodine for Antibacterial Activities. Wuhan Univ. *J. Nat. Sci.*10, 251-257.

Ward, P. D., Tippin, T. K. & Thakker. D. R. (2000). Enhancing paracellular permeability. *Pharm. Sci. Tech. Today,* 3, 346-358.

Wedmore, I., McManus, J. G., Pusateri, A. E. & Holcomb, J. B. (2006) A special report on the chitosan-based hemostatic dressing: experience in current combat operations. *J. Trauma,* 60, 655–658.

Weng, L. Romanov, A. Rooney, J. and Chen, W. (2008). Non-cytotoxic, in situ gelable hydrogels composed of N-carboxyethyl chitosan and oxidized dextran. *Biomat.,* 29, 3905–3913.

Winie, T. & Arof, A. K. (2004). Dielectric behavior and ac conductivity of LiCF3SO3 H-chitosan polymer films. *Ionics,* 10, 193-199.

Wongpanit, P., Sanchavanakit, N., Pavasant, P., Supaphol, P., Tokura, S. & Rujiravanit, R. (2005). Preparation and Characterization of Microwave-treated Carboxymethyl Chitin and Carboxymethyl Chitosan Films for Potential Use in Wound Care Application. *Macromol. Biosci.,* 5, 1001–1012.

Wu, Y., Dong, Y., Zhou, F., Ruan, Y., Wang, H. & Zhao, Y. (2003). Studies on the critical phase-transition behavior of cholesteric *N*-phthaloyl chitosan/dimethyl sulfoxide solutions by five techniques. *J. Appl. Polym. Sci.,* 90, 583-586.

Wu, Y., Seo, T., Maeda, S., Sasaki, T., Irie, S. & Sakurai, K. (2005). Circular dichroism induced by the helical conformations of acylated chitosan derivatives bearing cinnamate chromophores. *J. Polym. Sci. Polym. Phys.,* 43, 1354-1364.

Wu, Y., Seo, T., Sasaki, T., Irie, S. & Sakurai, K. (2006). Layered structures of hydrophobically modified chitosan derivatives. *Carbohydr. Polym.,* 63, 493-499.

Xiao, J. B., Chen, X. Q, & Yu H. Z. (2006). Adsorption of nucleotides on β-cyclodextrin derivative grafted chitosan. *Macromol. Res.,* 14, 443-448.

Xie, W. M., Xu, P. X. Wang, W. Lu, Q. (2001). Antioxidant activity of water-soluble chitosan derivatives. *Bioorg. Med. Chem. Lett.,* 11, 1699–1703.

Xie, W., Xu, P. X., Liu, Q., & Xue, J. (2002a). Graft-copolymerization of methylacrylic acid onto hydroxypropyl chitosan. *Polym. Bull.,* 49, 47–54.

Xie, W. M., Xu, P. X., Wang, W. & Liu, Q. (2002b). Preparation and antibacterial activity of a water-soluble chitosan derivative. *Carbohydr. Polym.,* 50, 35–40.

Xu, J., McCarthy, S. P., Gross, R. A. & Kaplan, D.L. (1996). Chitosan film acylation and Eefects on biodegradability. *Macromol.,* 29, 3436-3430.

Xu, X., Li, L., Zhou, J., Lu, S., Yang, J., Yin, X. & Ren, J. (2007). Preparation and characterization of *N*-succinyl-*N'*-octyl chitosan micelles as doxorubicin carriers for effective anti-tumor activity. *Colloids Surf. B: Biointerf.,* 55, 222-228.

Xu, Y., Du, Y., Huang, R. & Gao, L. (2003). Preparation and modification of *N*-(2-hydroxyl) propyl-3-trimethyl ammonium chitosan chloride nanoparticle as a protein carrier. *Biomat.,* 24, 5015-5022.

Xu, Z., Wanb, X., Zhang, W., Wang, Z., Peng, R., Tao, F., Cai, L., Li, Y., Jiang, Q. & Gao, R. (2009). Synthesis of biodegradable polycationic methoxy poly(ethylene glycol)–polyethylenimine–chitosan and its potential as gene carrier. *Carbohydr. Polym.,* 78, 46–53.

Yamaguchi, R., Arai, Y., Itoh, T. & Hirano, S. (1981) Preparation of partially *N*-succinylated chitosans and their cross-linked gels. *Carbohydr. Res.,* 88, 172-175.

Yan, C., Chen, D., Gu, J. & Lj, L. (2006) Synthesis of *N*-Succinyl-chitosan (Suc-Chi) and Preparation of Oxymatrine (OM)/*N*-Succinyl-chitosannanoparticles. *Chem. Res. Chin. Univ.,* 22, 589-592.

Yang, K. W., Li, X. R., Yang, Z. L., Li, P. Z., Wang, F. & Liu Y. (2009a). Novel polyion complex micelles for liver-targeted delivery of diammonium glycyrrhizinate: *In vitro* and *in vivo* characterization. *J. Biomed. Mater. Res. A.,* 88, 140–148.

Yang, H., Zhou, S. B. & Deng. X. M. (2005). Synthesis and Characterization of Chitosan-g-poly- (D, L-lactic acid) Copolymer. *Chin. Chem. Lett.,* 16, 123-126.

Yang, J. M. & Lin, H. T. (2004). Properties of chitosan containing PP-g-AA-g-NIPAAm bigraft nonwoven fabric for wound dressing. *J. Membr. Sci.,* 243, 1–7.

Yang, R., Yang, S.G., Shim, W. S., Cui, F., Cheng, G., Kim, I. W., Kim, D. D., Chung, S. J. & Shim, C. K. (2009b). Lung-specific delivery of paclitaxel by chitosan-modified PLGA nanoparticles via transient formation of microaggregates. *J. Pharm. Sci.,* 98, 970–984.

Yang, S., Tirmizi, S. A., Burns, A., Barney, A. A., & Risen, W. M. (1989). Chitaline materials: Soluble chitosan-polyaniline copolymers and their conductive doped forms. *Synth. Met.,* 32, 191–200.

Yang, X., Zhang, Q., Wang, Y., Chen, H., Zhang, H., Gao, F. & Liu, L. (2008). Self-aggregated nanoparticles from methoxy poly (ethylene glycol)-modified chitosan: synthesis, characterization; aggregation and methotrexate release in vitro. *Colloids Surf. B., Biointerfaces,* 61, 125–131.

Yang, Z.K. & Yuan, Y. (2001). Studies on the synthesis and properties of hydroxyl azacrown ether-grafted chitosan. *J. Appl. Polym. Sci.,* 82, 1838–1843.

Yazdani-Pedram, M., Retuert, J., & Quijada, R. (2000). Hydrogels based on modified chitosan, 1. Synthesis and swelling behavior of poly(acrylic acid) grafted chitosan, *Macromol. Chem. Phys.,* 201, 923–930.

Yin, L., Fei, L. L., Cui, F., Yin, C.T. & Yin, C. (2007). Superporous hydrogels containing poly(acrylic acid-*co*-acrylamide)/*O*-carboxymethyl chitosan interpenetrating polymer networks. *Biomaterials,* 28, 1258-1266.

Yoo, H. S., Lee, J. E., Chung, H., Kwon, I.,C. & Jeong, S.Y. (2005). Self-assembled nanoparticles containing hydrophobically modified glycol chitosan for gene delivery. *J. Control. Rel.* 103, 235-243.

Yoshida, H., Nishihara, H. & Kataoka, T. (1994). Adsorption of BSA on strongly basic chitosan: Equilibria. *Biotechnol. Bioeng.,* 43, 1087-1093.

Yoshioka, H., Nonaka, K., Fukuda, K., Kazama, S. (1995). Chitosan-derived polymer-surfactants and their micellar properties. *Biosci. Biotechnol. Biochem.,* 59, 1901-1904.

Yu, L. L., He, Y., Bin, L. & Yue, F. (2003). Study of radiation-induced graft copolymerization of butyl acrylate onto chitosan in acetic acid aqueous solution. *J. Appl. Polym. Sci.,* 90, 2855–2860.

Yuan, Q., Venkatasubramanian, R., Hein, S. & Misra R. D. (2008). A stimulus responsive magnetic nanoparticle drug carrier: magnetite encapsulated by chitosan-grafted-copolymer. *Acta Biomater.,* 4, 1024-37.

Yuan, Z. X., Sun, X., Gong, T., Ding, H., Fu, Y. & Zhang, Z. R. (2007). Randomly 50% N-acetylated low molecular weight chitosan as a novel renal targeting carrier, *J. Drug Target.,* 15, 269–278.

Zambito, Y., Uccello-Barretta, G., Zaino, C., Balzano, F. & Di Colo, G. (2006). Novel transmucosal absorption enhancers obtained by aminoalkylation of chitosan. *Eur. J. Pharm. Sci.*, 29, 460-469.

Zha, F., Lu, R. & Chang, Y. (2007). Preparation and Adsorption Property of Chitosan Derivative Bearing β- Cyclodextrin and Schiff-Base. *J. Macromol. Sci., Part A Pure Appl. Chem.*, 44, 413–415.

Zhang, C., Ping, Q., Zhang, H. & Shen, J. (2003). Preparation of N-alkyl-O-sulfate chitosan derivatives and micellar solubilization of taxol. *Carbohydr. Poly.*, 54, 167-141.

Zhang, H., Mardyani, S., Chan, W. C. & Kumacheva, E. (2006). Design of biocompatible chitosan microgels for targeted pH mediated intracellular release of cancer therapeutics. *Biomacromol.*, 7, 1568-1572.

Zhang, M. & Ren, H. X. (2007). Structural modification and application of chitosan. *J. Clin. Rehabil. Tissue Eng. Res.*, 11, 9817–9820.

Zhang, X., Wang, Y., & Yi, Y. (2004). Synthesis and characterization of grafting β-cyclodextrin with chitosan. *J. Appl. Polym. Sci.*, 94, 860-864

Zhang, X., Wu, Z., Gao, X., Shu, S., Huijie Z., Wang, Z. & Li C. (2009). Chitosan *bearing pendant cyclodextrin as a carrier for controlled protein release. Carbohydr. Polym.*, 77, 394–401.

Zhao, Z.P. Wang, Z. & Wang, S.C. (2003). Formation, charged characteristic and BSA adsorption behavior of carboxymethyl chitosan/PES composite MF membrane. *J. Membr. Sci.*, 217, 151-158.

Zheng, F., Shi, X. W., Yang, G. F., Gong, L. L., Yuan, H. Y., Cui, Y. J., Wang, Y., Du, Y. M. & Li, Y. (2007). Chitosan nanoparticle as gene therapy vector via gastrointestinal mucosa administration: results of an in vitro and in vivo study. *Life Sci.*, 80, 388–396.

Zheng, Y., Cai, Z., Song, X., Chen, Q., Bi, Y., Li , Y. & Hou, S. (2009). Preparation and characterization of folate conjugated N-trimethyl chitosan nanoparticles as protein carrier targeting folate receptor: in vitro studies. *J. Drug Target.*, 17, 294-303.

Zhu, A., Liu, J., & Ye, W. (2006). Effective loading and controlled release of camptothecin by *O*-carboxymethylchitosan aggregates. *Carbohydr. Polym.*, 63, 89-96.

Zhu, A. & Fan, N. (2005). Adhesion Dynamics, Morphology, and Organization of 3T3 Fibroblast on Chitosan and Its Derivative: The Effect of O- carboxymethylation. *Biomacromol.*, 6, 2607-2614.

Zhu, A., Jin, W., Yuan, L., Yang, G., Yu, H. & Wu, H. (2007). *O*-carboxymethylchitosan-based novel gatifloxacin delivery system. *Carbohydr. Polym.*, 68, 693-700.

Zhu, H.U., Ji, J., Lin, R. G., Gao, C. G., Feng, L. X. & Shen, J.C. (2002) Surface engineering of poly(D,L-lactic acid) by entrapment of chitosan-based derivatives for the promotion of chondrogenesis. *J. Biomed. Mater. Res.,* 62, 532-539.

Zong, Z., Kimura, Y., Takahashi, M. & Yamane, H. (2000). Characterization of chemical and solid state structures of acylated chitosans. *Polym.,* 41, 899-906.

INDEX